TEXTBOOK OF AGRICULTURAL FINANCE & CO-OPERATION

TEXTBOOK OF AGRICULTURAL FINANCE & CO-OPERATION

by

S G Thomas

Wordpen Academics

Publishing Science That Matter

ISBN: 978-93-6084-672-5

Published by : **WORDPEN ACADEMICS**
4736/23, Ansari Road, Darya Ganj,
New Delhi – 110 002
Phone: +91-011-35004336
E-mail: info@wordpenacademics.com

Exclusive Distributor: ***Bio-Green Books, New Delhi***

Preface

Agriculture, often regarded as the backbone of economies worldwide, relies not only on the fertility of the land and the expertise of the farmer but also on the financial mechanisms that support its growth. Agricultural finance and cooperation are crucial in ensuring the sustainability, efficiency, and profitability of agricultural enterprises, especially in a world that faces rapid demographic changes, technological advancements, and climate uncertainties.

This book, titled *Textbook of Agricultural Finance & Co-Operation*, provides an in-depth exploration of the principles, practices, and institutions involved in financing agriculture and fostering cooperation within the sector. Designed for students, researchers, policymakers, and practitioners, it aims to offer a comprehensive understanding of the economic and financial frameworks that drive agricultural development.

The initial chapters introduce the foundational concepts of agricultural finance, covering topics such as credit systems, risk management, investment analysis, and rural financial institutions. Readers are exposed to the various sources of agricultural financing, including formal institutions like banks and informal sources such as microfinance and cooperative credit societies.

The book then delves into the role of agricultural cooperatives in promoting collective action, improving market access, and increasing bargaining power for farmers. It provides a detailed analysis of the structure, governance, and operations of cooperatives, and their contribution to rural development and economic stability.

Subsequent sections examine the financial tools and strategies used by agricultural enterprises, including crop insurance, subsidies, and innovative financing models like impact investing and green bonds. Special attention is given to the role of public policy in supporting agricultural finance, as well as the challenges of financing smallholder agriculture in developing countries.

Real-world case studies from around the world illustrate how agricultural finance and cooperative models have been successfully implemented, offering valuable lessons for practitioners and policymakers.

Textbook of Agricultural Finance & Co-Operation serves as a vital resource for anyone involved in the agricultural sector, providing both theoretical insights and practical knowledge to address the financial challenges and opportunities within this essential industry.

S G Thomas

Contents

1

Introduction to Agricultural Finance

1.1 Meaning

Agricultural finance is the provision of multiple types of services dedicated to supporting both on- and off-farm agricultural activities and businesses including input provision, production, and distribution, wholesale, processing and marketing. The market demand for smallholder agriculture finance is US$450 billion, most of which is unmet. Without access to credit, most smallholders are restricted to farming practices that result in low levels of productivity. Although many new types of formal financial institutions have spread to the rural areas of developing countries, the majority of smallholders have limited or no access to them. Farm surveys often show that less than 1 in 10 smallholder farmers is eligible for a commercial loan from a bank or other financial organisation.

As agriculture sector contributes about 40% of worldwide employment and a 100% food production increase will be required in developing countries to feed the 2050 population, investment in agriculture sector is critical for driving global economic growth. The issues of food security, increased poverty in developing nations and overall imbalanced development of agriculture-dependent economies have highlighted the urgent need for development in that sector.

There is an ever increasing need to invest in agriculture due to a drastic rise in global population and changing dietary preferences of the growing middle class in emerging markets towards higher value agricultural products. In addition, climate risks increase the need for investments to make agriculture more resilient to such risks. Estimates suggest that demand for food will increase by 70% by 2050 and at least $80 billion annual investments will be needed to meet this demand, most of which needs to come from the private sector. Financial sector institutions in developing countries lend a disproportionately lower share of their loan portfolios to agriculture compared to agriculture sector's share of GDP.

Due to its inherent risky nature, the availability of formal financial services dedicated to the needs of smallholders is limited. The provision of credit in poor, rural areas where the risks are high and the returns low is unattractive to financial service providers. Typically such loans are very small and require careful attention and hence are very costly to service. Farmers, and especially women and young people, many who do not have formal ownership of their lands, have little to offer in the form of collateral. The traditional alternative of a money lender, can be extremely risky for a resource-poor farmer as the loans are charged at very high rates of interest contributing to increasing challenges of indebtedness.

On the other side, the growth and deepening of agriculture finance markets is constrained by a variety of factors which include:

- inadequate or ineffective policies,
- high transaction costs to reach remote rural populations,
- covariance of production, market, and price risks,
- absence of adequate instruments to manage risks,
- low levels of demand due to fragmentation and incipient development of value chains,
- lack of expertise of financial institutions in managing agricultural loan portfolios.

The development of agriculture requires financial services that can support: larger agriculture investments and agriculture-related infrastructure that require long-term funding (given that currently transportation and logistics costs are too high, especially for landlocked countries), a greater inclusion of youth and women in the sector, and advancements in technology (both in terms of mechanizing the agricultural processes and leveraging mobile phones and electronic payment platforms to enhance access and reduce transaction costs). An important challenge is to address systemic risks through insurance and other risk management mechanisms and lower operating costs in dealing with smallholder farmers.

Agriculture finance and agricultural insurance are strategically important for eradicating extreme poverty and boosting shared prosperity. Globally, there are an estimated 500 million smallholder farming households – representing 2.5 billion people – relying, to varying degrees, on agricultural production for their livelihoods. The benefits of our work include the following: growing income of farmers and agricultural SMEs through commercialization and access to better technologies, increasing resilience through climate smart production, risk diversification and access to financial tools, and smoothing the transition of non-commercial farmers out of agriculture and facilitating the consolidation of farms, assets and production (financing structural change).

The development of self-help and savings and loan groups has been one answer to this challenge; microfinance is yet another. Microfinance institutions (MFI's) develop their products for low-income households and individuals. Whilst MFIs are driving the growth and accessibility of rural, especially agricultural, financial services, their market share is still small. Most have focused on urban and peri-urban clients and rural areas with high population densities, leaving many rural areas and farmers still lacking access to credit.

Value chain financing is defined as the flows of funds to and among the various links within a value chain. The system allows borrowers to benefit through higher lending at better terms and obtain loans that reflect the cash flow pattern of their producing, processing or trading activities. The Caisse des Affaires Financières (CAF) in Isonga, Northern Rwanda, is example of successful value chain financing. The CAF introduced production and marketing loans, voucher systems and leases for transport that ultimately led to productivity increases of approximately 30% between 2007 and 2008. Furthermore, all farmers in Isonga now have bank accounts and the farmers' cooperative was able to acquire a truck to reduce transport costs.

1.1.1 Overview

From the perspective of winning trade wars, the United States has an insurmountable advantage in agriculture. However, sales of most U.S. agricultural products are not only unnecessary, they are morally wrong. While multiplying exporters GDP, these exports destroy native agriculture by usurping their local markets. The smaller level of circulating money within the economy (the sabotage of the multiplier factor) due to paying for imported food limits the development of, or even destroys industries in, other sectors of the economy.

One of the most sacred illusions of America is that its agriculture is above all reproach. Not only is the United States the "breadbasket of the world," but the developing world is somehow incapable of emulating America's productive farming methods. There is one thing Americans are sure about, without their food and generosity, much of the rest of the world would starve. Yet 40% of the developing world that was once plagued by severe food shortages—China, Guinea-Bissau and—until impoverished by embargoes—Cuba and North Korea produced and distributed the 2,300-to-2,400 calories per day required to sustain an adult. India has finally achieved and maintained self-sufficiency. Angola, Mozambique, and Nicaragua had also achieved self-sufficiency, but their economic infrastructures vere sabotaged by anti-government rebels organized, trained, and armed by U.S. or llied intelligence services.

The countries that are newly self-sufficient in food production have far less cultivable land per person than most of the countries still suffering from chronic food shortages. China, for example, has only .13 hectares of arable land per person; the former North Vietnam had 0.10; and North Korea (self sufficient before devastation by the Korean War and the embargo) has .07. Despite having more arable land per person, their neighbors are unable to feed themselves. Pakistan has .40 cultivable hectares per person; Bangladesh has .16; and Indonesia has .15 hectares.

The best-known example of a country that is continually faced with hunger is Bangladesh, where "two-thirds of the population suffers from protein and vitamin deficiencies." Yet the country exists on a fertile plain blessed with plenty of water, and "grows enough in grain alone to provide everyone in the country with at least 2,600 calories a day." It is obvious that nature has provided this country with the ability to feed more than the present population.

The reasons for such anomalies become clearer when one studies Africa and South America, the two continents with the hungriest populations. The United Nations Food and Agriculture Organization estimates that only 60% of the world's arable land is farmed. In Africa and South America, the figure averages 20%, and their grain yields are only one-half that of industrialized countries. Brazil, for example, is burdened with a large hunger problem, but, even without the destruction of more rainforests, it has cultivable acres per person. In Brazil, as well as most of South and Central America, one-half the acres being farmed—invariably the best land—currently grows crops for feeding cattle or for export. The masses are unable to feed themselves because their land is subtly monopolized. Brazil has ranches with up to 250,000 head of cattle (that one owned by the Rockefellers) which monopolizes land capable of feeding hundreds of millions of people. Latin Americans and Africans, despite rampant hunger, consume only a small percentage of their land's agricultural potential while a substantial share is exported.

The remaining hungry nations, mostly in Southeast Asia, have such large populations that the land's capacity to feed the people entails a much smaller margin of safety. Yet, if they controlled their land, these nations could also produce an adequate supply of food. China, probably the best example of rational land reform, now adequately feeds 1.3-billion people. But when the population was one-third w hat it is today and the land was monopolized, there were massive famines. Fifteen of the poorest countries in the world raise and export more agricultural products than they keep for their own use. Some of these countries, the exported crops, and the percentage of farmland thus removed from local consumption include: Guadeloupe—sugar, cocoa, and bananas, 66%; Martinique—bananas, coffee, cocoa, and sugar, 70%, and Barbados—sugar cane, 77%. Guatemala plants cotton for export in blocks of 50,000 acres.

These are all familiar developed world consumer items imported from these impoverished countries. In 1973, the United States imported 7% of its beef, much of it from the Dominican Republic and Central America. Costa Rica alone exported 60-million pounds to the United States in 1975, even though its own per-capita beef consumption dropped from 49 pounds per year in 1950, to 33 pounds in 1971. If Costa Ricans had not exported this increased production, their per-capita consumption would have been 3-times as high, or 98 pounds per year. While the United States imports all this beef, two-thirds of the grain it exports is used to feed livestock and much of the rest is distilled into liquors, both for elite consumption. In addition, it requires 40-cents' worth of imported oil to produce and transport every $1 worth of agricultural exports. "To produce and distribute 'just one can of corn containing 270 calories' consumes 2,790 calories of energy."

During 1992, U.S. food imports are estimated to have been $22-billion and exports $40-billion. Economists teach that there must be balanced trade and, from the perspective of maintaining the status quo, this may be true. However, the status quo reflects the unequal distribution of political and economic power in the world; the "geography of world hunger" is specifically the consequence of entire populations having lost control of their land, and thus their destiny. The impoverished world does not need America's, or Europe's, surplus food. They only need the right to control their own land, the right to industrial capital, and the right to grow their own food. Given those rights, they will not generally be hungry. However, because only the affluent have money to purchase this production, monopolization of land diverts the production of social wealth to those already well off. "The world can simply produce more than those who have money to pay for it can eat." The results are small well-cared-for elite groups, primarily in the developed world, and hunger for the dispossessed.

1.2 Scope

1.2.1 Introduction

Agricultural finance as the study of those financial intermediaries who provide loan funds to agriculture, and the financial markets in which these intermediaries obtain their funds. In fact, several agricultural economists identified a number of studies focusing on such additional topics as rural banking, insurance, income distribution, farm financial management, and taxation. Finally, the study of agricultural finance can be broaden even further to account for all economic and financial interfaces between agriculture and the rest of the macroeconomics, including the effects that changes in national economic policies have upon the economic performance of agriculture and the financial position of farm operator families.

Understanding financial concepts and the practical applications of finance is essential for anyone interested in pursuing a career in the agribusiness or agricultural production sectors. Many of the important managerial problems in agriculture involve finance. However, most agricultural production firms are significantly different from corporations, and more closely resemble small, owner-operated businesses. Management of farm business requires a wide range of information on physical and financial performance. Sometimes, however, much of the information needed is recorded only in the mind of the farm operator or in an informal, irregular kept ledger. There is also the tendency by some to judge the financial performance of a farm business by the amount of money he has in the bank. Each of these measures, however, provides a potentially misleading picture of the financial performance of the farm business and the financial strength of the farm operator.

1.2.2 Areas of Financial Knowledge

Farm Financial Management

The functions of financial management are traditionally defined to include the investment decision, financing decision, and the dividend decision. Together, these decisions largely determine the rate at which the farm business grows over time. Because most farm business is sole proprietorships, financial management in agriculture encompasses the proprietor withdrawal decision (i.e. the withdrawal of funds to finance personal consumption and nonfarm investments) instead of the dividend decision.

The investment, financing, and proprietor withdrawal decisions are not made independently. For example, the decision to invest in a new piece of farm equipment cannot be made independently of the financing or withdrawal decisions. Furthermore, the financing decision is directly influenced by the farm operator's decision to withdraw part of his current net farm income to finance personal consumption and nonfarm investment.

Effective investment, financing, and withdrawal decision making requires a comprehensive farm financial accounting system. For example, an accounting system is needed to help identify the level and timing of financing needed to facilitate the farm operator's current production and marketing plans. Such an accounting system should also provide information pertaining to the farm operator's present financial position and the efficiency and profitability of his current operations.

Financial Accounting System

An accurate and comprehensive set of financial statements is essential to sound financial management. Such an accounting system should at a minimum include a balance sheet, an income statement, a statement of change in owner equity, and a cash flow statement. These statements can provide an informational input to the production, marketing, investment, and financing decisions that must be made annually by the farm operator. For example, the information contained in these statements can be used to evaluate the annual performance of the farm operator's business. This could be done by comparing the profitability and efficiency ratios computed for his business in the current year with similar ratios he has achieved in previous periods or with other farm businesses having similar types of operations.

Organization and Growth of the Farm

To maximize the profits generated by the farm business in the current period, the operator must be able to identify the appropriate combination of products to produce and inputs to use. In the longer run, the farm operator must also be able to determine how large the farm should be and understand what factors determine how fast he can reach this size. Some of the externally imposed factors that can constrain the rate of growth of the farm are external credit rationing, taxes, and government regulations. The farm operator can also internally constrain the rate of growth of the firm by rationing his use of credit or by withdrawing a larger part of net farm income to finance personal consumption expenditures or non-farm investments.

Investment and Financing

The concept of the time value of money underlines much of investment theory. For example, the present value of the annual net cash flows generated by an investment over its economic life must be taken into account when choosing between alternative investment projects.

The presence of business and financial risk must also be taken into account when evaluating alternative investment projects with differing level of risk. The farm operators' risk-return preferences (i.e., whether they are averse to risk, indifferent toward risk, or prefer risk) can have an effect on the ranking and acceptance of alternative investment projects available to them. To reduce their exposure to risk, farms operators can adopt one of several different strategies. These strategies include product diversification, sequential marketing of farm output, insurance, and forward contracting using either futures market contracts or cash forward contracts.

Legal Considerations

There are several forms of legal documents that farm operators should be familiar with when borrowing or leasing. These contracts document the rights and obligations of the borrower and lender, or lessor and lessee. It is important that the borrower and lessee understand the performance expected of them as set forth in these contracts.

Lenders, on the other hand, are concerned with the security of loans to protect themselves in the event that the borrower defaults on a loan. In some cases, the lender will accept a cosigner of a note. In other cases, however, a security agreement documenting the property the lender has a security interest in is used. A security interest gives the lender rights to those assets the borrower has pledged as collateral. The lender is also concerned about the borrower receiving a clear title to the property being purchased, particularly in the case of real estate.

There also legal considerations unique to leasing. For example, a legally enforceable lease must contain the names of the parties involved, a description of the property, the time period involved, and the timing and form of compensation. Farm land is typically leased on a cash rent, crop-share, or livestock-share basis. Machinery, on the other hand, is leased either through an operating lease or a financial lease.

Taxation

To maximize after tax profits, farm operators must have a working knowledge of tax laws and the options available to them. Gains or losses associated with the sale of specific types of assets and the effects of minimum tax and self-employment tax will be discussed. The differences in the tax laws as they apply to sole proprietorships, partnerships, or corporations are also identified.

Financial Intermediaries Serving Agriculture

The financial intermediaries serving agriculture provide an important service by channeling funds from savers to borrowers in the amounts necessary to finance production expenses and capital expenditures. The funds used by these financial intermediaries come from variety of sources. These funds are then made available to agriculture under a variety of terms.

1.2.3 Scope in India

Farm finance has become an important input due to the advent of capital intensive agricu1tural technologies. Farmers require capital in order to enhance the productivities of various farm resources. Indian agriculture, in general, is characterized by low and uncertain returns. In order to break the vicious cycle of low returns → low savings → low investment → low returns, provision of external finance to farmers becomes inevitable.

Organized and unorganized credit agencies in rural area provide credit for both development and consumption purposes. Provision of credit by these agencies involved many obstacles to both bankers and borrowers due to differences in banking system followed by bankers, socio-economic conditions of borrowers and infra - structural facilities and institutional support offered to the borrowers. Besides, the government also frequently changes its agricultural credit policies regarding institutional credit set-up, credit rationing, rates of interest, subsidy and the functioning of markets and other developmental agencies which would influence the extent of credit available to farmer-borrower. All these factors, therefore, ultimately have a bearing on farm returns. Hence, problems regarding capital could be well understood, if one could realise the theoretical basis of agricultural credit system in India, bottlenecks faced by bankers and borrowers, and the governments' efforts in solving the problems involved in the agricultural credit system in India.

Importance

Credit is essential for agricultural development and also for the development of the economy as a whole. The agricultural finance is required for the following reasons:

- ❖ The scope for extensive agriculture in India is limited. Therefore, increase in agricultural production is possible only by intensification and diversification of farming. Intensive agriculture needs huge capita.
- ❖ Extreme inequalities exist in the distribution of operational holdings and operational area. 74.5 per cent of the total number of farm households which own less than 2 hectares operate only 26.2 per cent of the total operated area whereas only 2.4 per cent of total number of farm households which own more than 10 hectares each operate 23 percent of the total operated area in 1980-81 (In India, there were 88.883 million farm households which operated 163.797 million hectares in 1980-81). The purchasing power of these small and marginal farmers is limited to their subsistence farming. Hence, they have to depend on the external financial assistance to use the costlier (modern) inputs.
- ❖ Farmers economic condition is subject to frequent onslaught of flood, drought, famine etc. Therefore, either the continuance of cultivation of crops or making improvements on the farms depends on the nature and availability of finance.
- ❖ In recent years, more area is brought under irrigation which in turn would increase the use of inputs like fertilizer and plant protection chemicals. In order to accomplish this, external finance is needed.
- ❖ In order to sustain the development of agro-based industries, there should be a substantial increase in the supply of raw materials needed for such industries. Therefore, for the development of farm sector, a constant flow of credit is essential and it would enhance overall growth of the economy.
- ❖ In agriculture, fixed capital is locked up in permanent investments like land, well, buildings, etc. Moreover, it takes a long time to get returns from farm. Hence, farmers need finance to continue their farm operations.
- ❖ The weaker sections of the farming community should be motivated to participate in development programmes by giving financial assistance to acquire productive assets.

- Small and marginal farmer's are trapped in the vicious cycle of poverty i.e., low returns → low saving → low investment → low return. To break this cycle, credit has to be injected in agricultural sector.

1.3 Significance

1.3.1 Introduction

Professional moneylenders were the only source of credit to agriculture till 1935. They use to charge unduly high rates of interest and follow serious practices while giving loans and recovering them. As a result, farmers were heavily burdened with debts and many of them perpetuated debts. There were widespread discontents among farmers against these practices and there were instances of riots also.

With the passing of Reserve Bank of India Act 1934, District Central Co-op. Banks Act and Land Development Banks Act, agricultural credit received impetus and there were improvements in agricultural credit. A powerful alternative agency came into being. Large-scale credit became available with reasonable rates of interest at easy terms, both in terms of granting loans and recovery of them. Both the co-operative banks advance credit mostly to agriculture. First bank advances short-term and medium term loans while the second bank advances long-term loans. The Reserve Bank of India as the Central bank of the country took lead in making credit available to agriculture through these banks by laying down suitable policies.

Although the co-operative banks started financing agriculture with their establishments in 1930's real impetons was received only after Independence when suitable legislation were passed and policies were formulated. Thereafter, bank credit to agriculture made phenomenal progress by opening branches in rural areas and attracting deposits.

Till 14 major commercial banks were nationalized in 1969, co-operative banks were the main institutional agencies providing finance to agriculture. After nationalization, it was made mandatory for these banks to provide finance to agriculture as a priority sector. These banks undertook special programs of branch expansion and created a network of banking services through out the country and started financing agriculture on large scale. Thus agriculture credit acquired multi-agency dimension. Development and adoption of new technologies and availability of finance go hand in hand. In bringing "Green Revolution", "White Revolution" and now "Yellow Revolution" finance has played a crucial role. Now the agriculture credit, through multi agency approach has come to stay.

The procedures and amount of loans for various purposes have been standardized. Among the various purposes "Crop loans" (Short-term loan) has the major share. In addition, farmers get loans for purchase of electric motor with pump, tractor and other machinery, digging wells or boring wells, installation of pipe lines, drip irrigation, planting fruit orchards, purchase of dairy animals and feeds/fodder for them, poultry, sheep/goat keeping and for many other allied enterprises.

The Agriculture sector continues to play a central role in the country economy. It is the second largest sector, accounting for 20.9 percent of the Gross Domestic Product (GDP) in 2014-15 and is a source of livelihood of 43.5 percent of rural population. Increased agricultural production and high crops yield is essential for food security which make the farming systems less vulnerable to climate change. To make agriculture more effective in supporting sustainable higher economic growth trajectory and reducing poverty, a policy framework needs to be anchored coupled with favorable socio political climate, adequate governance, and sound macroeconomic fundamentals. The prime focus of the government is on high value agriculture including horticulture, livestock and fisheries. Concerted efforts are being made to improve farm level practices and developing linkages of farmers with markets and industry based on new technologies, ideas and future pathways for sustainable growth of agro industry. Nearly 62 percent of the country's population resides in rural areas, and is directly or indirectly linked with agriculture for their livelihood. The Agriculture sector's strong linkages with the rest of the economy are not fully captured in the statistics. While on the one hand, the sector is a primary supplier of raw materials to downstream industry, contributing substantially to Pakistan's exports, on the other, it is a large market for industrial products such as fertilizer, pesticides, tractors and agricultural implements.

The potential role for agriculture in development is to reduce poverty and drive growth for countries whose economies are agriculture-based. Growing population size requires agriculture growth compatible to meet required level of food. The change in consumption pattern with a change in per capita income level requires more proteins containing diet. The transition of agriculture from traditional to modern farming techniques is based on adequate availability of inputs like certified seeds, balanced use of fertilizers, mechanization, and agricultural finance. Agricultural finance plays an important role in enhancing the agricultural productivity in developing countries. Finance is the back bone for any business, more so for agriculture which has traditionally been a nonmonetary activity for the rural population. Rural credit, though not a direct tool of production, can help break the vicious circle of 'groweat-grow' by removing financial constraints and accelerating the adoption of new technologies. Credit facilities are thus the integral part of the process of commercialization of the rural economy. The introduction of easy and cheap credit is the quickest way to give boost to the agricultural production. Therefore, it was the prime policy of all successive governments to meet the credit requirements of the farming community of the country.

Credit is an important tool for getting the inputs in time increasing thereby the productivity of the farms particularly those of small ones. The current study was designed to investigate the problems faced by the farmers while getting the loan. It was found that the small farmers faced a lot of problems in getting and returning the loan which must be removed to get better results and hence improving the quality and quantity of the agricultural products. The use of credit facilities would therefore translate to higher resource employment and capacity utilization, increased output and income, and reduce poverty in the rural economy, especially among the farmers and be helpful to increase the food production which would lead to an improvement in the welfare of the farmers and consequently a reduction in their poverty and food insecurity levels.

1.3.2 Sources

The agricultural credit/finance system of the country consists of informal and formal sources of credit supply. The informal sources include friends, relatives, commission agents, traders and private moneylenders etc. Presently, the formal credit sources are comprised of financial institutions known as Agricultural Development Bank, Commercial Banks, and Federal Bank for Cooperatives. Recently some non-government organizations are also advancing agricultural credit to the rural communities.

Traditionally, friends and relatives, village shopkeepers, traders, commission agents, etc. have remained a major source of agricultural credit. These sources generally lend for short periods and charge an exorbitant rate of interest. Such loans are given to tide over bad periods and as such are meant for consumption purposes. Loans are also made available for buying seasonal inputs where cash is essentially required. These sources are both inadequate and non dependable. No comprehensive data are available on the amount of credit advanced by informal credit sources. As such it is difficult to find solid evidence regarding the relative share of these sources in the total credit supply. However, few reports and some other rough estimates show that the formal credit sources have been able to meet only 50 percent of the total credit requirements of the farm sector and it may be assumed that the rest are met by informal sources of credit. These sources play a vital role in our rural economy, especially for meeting the consumption requirements of thousands of small farmers along with fulfilling the need for irrigation water, hiring of farm power, cost of fuel etc. Mark-up tends to vary from product to product and also by type of borrower as well as lender. On the average, the mark-up is estimated to be 25 percent. In the case of fertilizer it is 29 percent while for the pesticides it comes around 35 percent. High mark-up charged on pesticides sold on credit is mostly due to rather unholy alliance between the manufacturers and traders/dealers. The case studies suggest that pure money lending is on the rise with quite exploitative interest rates ranging from 48 percent to 120 percent per year.

Agricultural finance is an essential part of the process of uplifting of agriculture and rural economy. Agriculture finance is systematically institutionalized for small farmers, agricultural development cannot be materialized. Due to small holdings, low crop yields and small income there is very little saving among the majority of the farmers of the country. Therefore, it is need of time that financing agencies come up to help them in undertaking the improved farm practices.

1.3.3 Features

As a farmer, it can be difficult to purchase the equipment and machinery you need. The costs can be exorbitant and can eat into capital that is much needed for other necessities. You may not be aware, but there is a solution to this in the form of agricultural finance. Outlined below is the importance of agricultural finance.

It will allow you to purchase new inputs

Farmers need to purchase new inputs, such as seeds, fertilizers, pesticides, irrigation water and more. Agricultural finance can help to make these purchases easier for farmers. If the seed of a high yielding crop is readily available for farmers, then the productivity of the farm is improved.

Smaller farms may not have the need for agricultural finance for items such as seeds or pesticides but larger farms may need help with bulk purchases of these items. Seeds, fertilizers and irrigation water can prove to be a highly expensive continuing need which agricultural finance can help to meet.

You can permanently improve your land

Having agricultural finance means that you are able to make permanent improvements on your land, such as sinking wells, rotation of crops and even land reclamation. If you were to attempt such improvements without finance, you may end up spending more money than you are able.

Finance for agricultural improvements means that you can create the perfect working farm to improve the productivity of your workers and the output of your land. Being able to rotate crops effectively will help significantly with better crop production and sustainability of farmland, and sinking wells will add value to your land should you decide to sell it.

You can cover land costs

If you are looking to buy new farmland as a budding farmer or simply increase the amount of land you already have, then agricultural finance can help cover the land costs you may incur. The land you need will depend on the type of farming you are planning on doing.

In order to apply for finance for land, you will need to calculate how much land you need and what kind of land you are looking for. Once you have your loan approved, you will be able to move forward with your endeavour. Buying land with your own money may not be feasible as a start-up farm, which is why finance is a good option.

It allows you to refinance an older loan

If you already have an agricultural loan, you may want to look into refinancing. This entails taking out a new, lower-interest loan and using this to pay off the old, higher-interest one. You will still have a loan but will be saving money due to the lower interest.

This decision will only make sense if the interest savings outweigh the refinancing costs. Once you have discussed your options with a financial advisor, you can look into refinancing. It will allow you to use your money for other essentials that you may not have been able to while repaying the older, expensive loan.

Marketing of products

If you are a working farm that sells its produce or livestock, you will need to successfully market your products. This can involve significant costs including websites, logos, focused ad campaigns, PR and marketing costs.

For those who are unfamiliar with marketing, you may need to speak to a consultant which will also incur consulting costs. You can use your agricultural finance as part of your marketing funds to boost your visibility for retail customers and for private customers too. Be sure to examine every aspect of any marketing campaign before making any final decisions.

You are better equipped for a crisis

Farming can be a difficult and fickle business. You are never able to predict what will happen to your crops or livestock, and are at the mercy of customers and competitors. Some farming is seasonal, which means you may only earn money during certain times of the year.

An agricultural loan can be used to protect yourself during the various ups and downs of your business. You can also use it for operational costs as well as costs that occur from damages. It is better to be prepared for every eventuality, which is why having agricultural finance is important to all working farms.

Final verdict

There are many benefits to taking out finance for agricultural pursuits. Your land improvement costs will go down, as well as being able to refinance your current loan. Farmland purchases are notoriously expensive, which is why it is best to take out finance to cover the costs. If you are concerned about not being able to make it until the end of the year due to seasonal crops, taking out finance may be an affordable solution.

1.4 Agricultural Economic Policy

The analysis of the effects of price and price distorting policies on the agricultural sector has received a great deal of attention in the economic literature pertaining to Sub–Saharan Africa. This literature suggests that inappropriate price and price–related policies have been the key impediment to agricultural growth and development in most Sub–Saharan countries. This finding appears to have had a significant influence on the design of structural adjustment programs (SAPs) in which getting the structure of relative prices right was considered to be the leading operational objective. Liberalization of input, output and service prices, marketing and trade, privatization of most public enterprises, devaluation of local currencies and disengagement of the state from most support services were primarily meant to achieve this objective.

With respect to privatization of parastatals, policy reforms implicitly assumed that the private sector will take over all the functions performed by these parastatals in a more efficient and cost–effective manner. It is, however, important to observe that the advocates of reforms appear to have paid little attention to some key questions related to the privatization process. Little attempt was made to identify functions that are best performed by government agencies and those that are best handled by the private sector or to assess the private sector base in each country concerned. The failure to examine these and other related key questions has made it difficult for the designers of the structural adjustment reforms to propose appropriate policy measures and actions that could help strengthen and foster the development of the private sector in order to enable it to effectively handle various functions that were previously carried out by parastatals in the economy. In addition, these policy reforms also failed to be specific about the timing and sequencing of the privatization process in order to avoid the disruption of agricultural and other economic activities.

Furthermore, structural adjustment programs appear to have paid limited attention to reforming government public goods, institutional and human capital development policies. These policies are even more central to the long–term development process of the economy than the stabilization and adjustment policies that have been the focus of the structural adjustment programs in Sub–Saharan Africa. Since the main objective of SAP was to get the structure of relative prices right, a question one may wish to address is did these policy reforms achieve this operational objective. What is clear here is that limited attention was paid to the impact of external factors on the structure of relative prices. The whole SAP package seems to have over–estimated the magnitude of leverage national governments in Sub–Saharan Africa have to affect the domestic agricultural terms of trade and hence the growth of agricultural output and income.

In fact, the failure of structural adjustment programs to improve the economic situation in Sub–Saharan Africa is now recognized both explicitly and implicitly by the opponents and designers of this policy reform package. As one of the most seasoned African economists put it "the continent" has the dubious distinction of being the only developing region of the world that experiences zero average per capita growth over the last thirty years, including negative growth rates over the last two decades. Embedded in the last two decades of the negative per capita growth is more than ten years of implementation of structural adjustment programs in most of Sub–Saharan Africa. After such a period of time, it is now appropriate to assess the impact these policy reforms have had on the agricultural sector.

This chapter attempts to evaluate the impact of economic policies undertaken since the 1970s including structural adjustment policy reforms on domestic terms of trade of major tradable agricultural commodities and in so doing to establish the extent of leverage governments in Sub–Saharan Africa have to affect the structure of relative prices facing their economies. More specifically, the examines the movements of domestic agricultural terms of trade and its major components in the light of economic policy reforms initiated in Sub–Saharan Africa; assesses the impact of external factors and domestic policy variables on the real exchange rate, one of the key components of the domestic terms of trade of tradable commodities; and finally estimates the contribution of external factors and domestic policies to change in the domestic terms of trade of agricultural tradables.

The West African countries, Niger and Senegal, were selected for the study. As can be seen, this sample of countries is made up of one coastal non–sahelian country, one coastal sahelian country (Senegal), and one landlocked sahelian country. These countries are all members of the West African Economic Monetary Union. They have the same currency that is linked to the French franc by a fixed nominal exchange rate. This exchange rate which was pegged at 50 CFA francs for one French franc since the 40's was changed on January 11, 1994 to 100 CFA francs for one French franc. This policy change has been one of the major macro–economic policy reforms initiated collectively by these and other countries of the Union.

The following crops are selected for the study, cocoa, cotton and rice in Côte d'Ivoire and, groundnut, cotton and rice in Senegal and Niger. It should also be observed that in Niger cotton and groundnut are the major export crops, while rice is the main importable food crop. Despite its decline, groundnut still remains a leading export crop in Senegal, while cotton, a relatively new export crop, presents significant potential for growth. With respect to rice in Senegal, it is not only the most important importable commodity, but a leading food crop.

The agricultural sector has played and continues to play a leading role in Côte d'Ivoire's economic development and the sector's rapid growth in the 1960s was the basis of what has been called the "Ivorian economic miracle." Although the

Ivorian economy is relatively diversified, it remains dependent on agriculture, which contributes almost one-half of GDP and employs about 54 percent of the economically active population. Two major crops dominate this sector: coffee and cocoa.

Coffee contributes about 50 percent of the country's export revenue and Côte d'Ivoire was the second largest African producer in 1992 after Ethiopia. Coffee farms are 98 percent small in size and the robust type dominates production. Cocoa production doubled in the 1970s and Côte d'Ivoire became the world's largest producer in 1977-1978 when its production over-took Ghana's. A state marketing agency, the Caisse de Stabilization et de Soutien des Price des Productions Agricoles (CAISSTAB) traditionally purchased all coffee and cocoa production before its privatization in early 1990s. Furthermore, Côte d'Ivoire has managed to diversify production in rice, cotton and rubber and the country has achieved self-sufficiency in almost all the food crops, except rice.

During 1965-73, sub-period was marked by high rate of growth of agricultural output and gross domestic product. The government development strategy was based on external borrowing and extraction of agricultural surplus to increase investments in basic infrastructure and other sectors. The CAISSTAB played a major role in this area by paying producer prices which were below the international level.

The 1974-78 sub-period was marked by various policy developments. The most notable 8development was the proliferation of state-owned companies. The sub-period was also characterized by an unprecedented boom in primary commodity prices and subsequent increasing export earnings. The latter helped the country to enlarge its industrial base in the areas of energy, agro-based industries (sugar, palm oil, coffee and cocoa processing). In the agricultural domain, the country decided to diversify production into new crops (cotton, rubber, sugar) and extending export and food crops out of the cocoa belt. The 1973-74 oil shock did not force to reduce investment expenditure, instead the country resorted to heavy external borrowing in order to maintain high investment rate. At the end of this sub-period, the country started to experience difficulties to service its heavy external debt.

This sub-period was characterized economic and financial crisis which prompted the country to apply austerity measures. Incapacity to reimburse external debt was one of the factors which led to negotiate an economic recovery program in 1980 with the IMF. This helped the country to obtain its first series of debt rescheduling in 1983, 1984 and 1986. But new adverse developments in export revenues led to further deterioration of economic and financial situations. The occurrence of drought in 1983 and 1984 caused production reduction in agriculture with coffee being one of the most affected. Furthermore, a prolonged collapse of primary commodity prices on international markets led to a severe decline in export revenues. Despite this, it was decided to increase producer prices in mid-1980s and the replanting program was maintained. Faced by shrinking export revenues, the government reacted to world price collapse by attempting to stockpile cocoa production. This unsuccessful attempt led to further erosion of the country's financial liquidities and subsequent increase in domestic and external debt arrears. Therefore, the country has no choice but to adopt the structural adjustment program.

The 1989–93 sub-period was marked by government's efforts to apply structural adjustment programs and a reverse producers' price policy was adopted. The structural adjustment policies focused on extended financial stabilization measures and structural reforms in order to create a competitive environment for the country. Stabilization measures were reflected in public investment and public employment reduction. Other structural measures included, inter alia, the liberalization of external trade, reduction of corporate tax and reduction of government role in the productive sectors.

With regard to producer prices, further deterioration in world prices for primary commodities led to reducing coffee and cocoa producer prices by half, while the government encouraged producers to increase food crop production in areas where self-sufficient was yet to be achieved. Despite these measures, improvement in leading economic indicators including the level of the country's competitiveness was rather limited. This led the country to accept along with other member countries in the WAMEU to devalue their commonest currency the CFA in January 1st, 1994.

1.5 Agriculture and Rural Development

Stabilization and Structural Adjustment Policies (SAPs) have been pursued in Sub-Saharan Africa (SSA) since the early 1980s to restore internal and external equilibria by controlling aggregate demand, and by liberalizing markets to reinstate the role of the price mechanism in efficient resource allocation.

Within the SAP framework, financial sector reform/restructuring consisted of:

- Liberalization of nominal interest rates from a fixed controlled regime to a market-determined regime, to mobilise savings and allocate loanable funds to projects with the highest returns on investment;
- Removal of credit allocation quotas and interest rate ceilings to particular sectors, including agriculture, to allow financial services to flow to the most productive uses;
- Privatizing parastatal banks and other financial institutions, to remove administrative inefficiencies and bankruptcy due to government bureaucracy, political interference and rent-seeking activities; and
- Tightening of banks supervision to ensure profitability by enforcing the recovery of non-performing assets, and solvency by increasing the equity base and tightening prudential regulations.

Implemented as such, financial sector restructuring was expected to promote long-term economic development. Since the relationship between financial sector restructuring and economic development has been examined by five papers sponsored under the research network, "African Perspectives on Structural Adjustment," it will not be re-examined here. In this financial sector restructuring is affecting long-term agricultural development and off-farm activities in the rural sector, which is the largest single sector in the SSA economies. An evaluation of the World Bank's early experience over the period by Jayarajah and Branson reveals that the approach to financial sector restructuring is practically independent of agricultural sector reform. Within agriculture, the reforms focused upon: decontrol of producer prices, removal of the implicit tax on agricultural exports by liberalizing the nominal exchange rate; dismantling state marketing monopolies; and inflation control to improve the rural/urban terms of trade.

For financial sector restructuring, the reforms concentrated on: removing financial repression; rehabilitating the formal banking system; and getting capital markets started. While it was acknowledged by Jayarajah that "Financial sector reform in many developing countries is more a process of development than reform" this development element was not explicitly incorporated into the restructuring of the financial sector in order to meet the development needs of the single largest real sector of the SSA economies.

In the World Bank Report covering the most recent period, Adjustment in Africa: Reforms, Results, and the Road Ahead, agriculture and financial sector restructuring do not show any change in the direction of the World Bank's policy thrust. Whereas the need for adequate availability of credit, to sustain the SAPs price-based incentives to promote agricultural sector growth, is acknowledged, in the second volume Adjustment in Africa: Lessons From Country Case Studies, the how to provide the needed credit is neither elaborated nor linked to financial sector restructuring.

Meanwhile, Montiel's review covering seven SSA countries: Ghana, Gambia, Kenya, Nigeria, Malawi, Tanzania and Uganda, is sounding a discordant note. The experience to-date in these countries, according to the review, suggests that the process of financial sector restructuring has not achieved much success: " financial liberalization, though recently undertaken in various countries of the region, has been tentative, and has thus far not appeared to be very successful." The key researchable issue, therefore, is whether the way in which this process is being conducted can promote long-term economic development, while omitting the largest rural sector, that is, "The issue confronting policy makers in sub-Saharan Africa is not whether further financial liberalization is desirable, but rather when and how it should be brought about." The purpose of this is to examine this issue systematically with respect to agriculture and off-farm activities in the rural sector.

Uganda is chosen because it is lately considered the most successful implementer of the SAPs on the SSA continent by the Institutions. The macro–economic success indicators over the period include: a real GDP growth rate of 5–6% p.a., a slashed inflation from over 300% to just around 10% p.a., and a parallel market exchange rate premium of over 300% reduced to zero, for example. The Uganda economy is basically rural. Agriculture provides over 60% of rural household income, with the rest coming from a wide range of off–farm activities: brewing local beer, fishing, handicraft, brick and charcoal making, construction and maintenance of owned–occupied dwellings etc. The Agricultural sector employs around 80% of the entire active labour force: this contributed 45% to real GDP in constant 1991 prices in 1996/97, and 90% to exports revenue. Agriculture is 90% organised by scattered small holders; estates are confined to tea and sugar production.

Amid the successful macroeconomic achievements, Uganda started financial sector restructuring in 1993 to remove financial repression, strengthen the banking system, and open a capital market. Yet the review of this financial sector restructuring program by the Republic of Uganda, Agricultural Policy Committee states that this program has no special action to help development of the rural financial markest. In fact, the rationalization of the branch network and liberalization of credit allocation can have a negative impact on the role of the banks in rural credit. The liberalization of credit allocation has had the tendency of crowding out agriculture from the banks' loan portfolio.

Faced with this brewing crisis by 1996, three years after the start of restructuring the financial sector, a number of proposals and studies are on the table debating how to provide rural financial services. "The challenge facing the banking system is therefore how to expand and broaden the financial intermediation in rural areas since the agricultural sector is crucial for economic development. This is a serious ongoing debate addressing a major defect in the design of the SAPs.

2

Agricultural Credit

2.1 Introduction

Agriculture plays a vital role in the overall development of the rural economy in particular and in the national economy in general in developing countries like India. Rapid and sustained agricultural development facilitates the process of economic growth. Agriculture development implies increased production and productivity of crops, generation of employment opportunities and thereby improved standard of living of the peasants. Moreover, the agricultural sector accelerates the overall economic development.

As per the promotion of agricultural benefits to mankind, progress of agriculture and the progress of civilisation go hand in hand." The industries for the supply of raw materials depend on agriculture. It is accepted that industrialisation can follow the sound heels of agriculture. If agriculture stagnates, it will act as brake on industrial expansion and halt real growth. The agriculture sector would alone act as a catalyst in breaking the vicious circle of poverty. The Government has given more importance to agriculture in the process of economic development in the country.

Credit is obtaining control over the use of money at the present time in exchange for a promise to repay it at some future time. It is also defined as a device for facilitating the temporary transfer of purchasing power from those who have surpluses of it to those who are in need of it. Thus, credit involves a temporary transfer of wealth. Agricultural Credit is the amount of investment funds made available for agricultural production from resources outside the farm sector.

Agricultural Finance is considered as separate field of study dealing with lending and borrowing by organizations and farmers. As referred, agricultural finance as the means of acquiring and control of assets, ownership by cash purchase or borrowing or leasing or custom-hiring.

Warren F.Lee et al defined Agricultural Finance as the economic study of the acquisition and use of capital in agriculture. It deals with the supply of and demand for funds in the agricultural sector of an economy. According to William G. Murray, agricultural finance is the economic study of borrowing of funds by farmers; of the organization and operation of farm lending agencies; and of society's interest in credit for agriculture.

Agricultural credit is considered as one of the most basic inputs for conducting all agricultural development programmes. In India there is an immense need for proper agricultural credit as Indian farmers are very poor. From the very beginning the prime source of agricultural credit in India was moneylenders.

After independence the Government adopted the institutional credit approach through various agencies like co-operatives, commercial banks, regional rural banks etc. to provide adequate credit to farmers, at a cheaper rate of interest. Moreover, with growing modernisation of agriculture during post-green revolution period the requirement of agricultural credit has increased further in recent years.

2.1.1 Farm Finance

Farm Finance is a branch of agricultural economics which deals with the provision and management of services of financial resources related to the individual farm units. Farm finance can also be defined as the amount of funds obtained from off-farm sources for use on the farm, repayable in future with an interest agreed to either explicitly or implicitly.

- ❖ is not meant merely for more production but also to raise the productivity of farm resources;
- ❖ not a mere loan or advance , but it is an instrument to promote the well being of the farming community;
- ❖ is not just a science to manage the money, but is an applied science of allocating scarce resources to derive optimum output; and
- ❖ not a mere social obligation on the society; but it is a lever with backward and forward linkages to the economic development both at toe micro and macro level.

At macro level, farm finance may be defined as the study of impact of finance (extended to the farmers by the intermediaries) on agricultural sector and also on the economy as a whole. At micro level, farm finance may be defined as the study of these intermediaries who extend finance to the farming sector and obtain their loanable funds from financial markets. Thus, farm finance should have the following features:

- ❖ finance should be extended to farmers for farm activities;

- finance should stimulate tie productivities of farm resources resulting in higher economic returns for the investment;
- finance should promote economic development of farm households;
- finance should be provided by an external agency for strengthening the backward and forward linkages with country's economic development.

2.2 Significance

The word credit is derived from the Latin word "Credire" which means to believe or to have a trust or have a faith or confidence. Credit is a means of obtaining resources at a certain period of time, with an obligation to repay it at subsequent period in accordance with the terms and conditions of the credit obtained". Credit is an important input in the production process. It is beneficial not only to the person who avails it of, but also to the person who parts away with it. It is a financial asset of the bank. It helps both the bank and the borrower in strengthening their financial status.

In a broader sense, credit is not only the life-blood of modem economic system but also the magic word of socio-economic transformation and a lever of development. Credit is one of the most crucial but scarce inputs used in agriculture. Farm credit is an important instrument, which has been used to increase agricultural productivity. Credit is the most important input of modern farming. The importance of credit has many facets in the Indian economy. It is the strategic input and integrated factor, which stimulates the growth impulses of the rural sector. The limited capital availability has been acknowledged as the most limiting factor in the modernisation of agriculture.

The credit has assumed a place of cardinal importance in India in the context of agricultural development owing to the gap of agriculturists requirements for agricultural operations and the ability to finance the same from their own funds. Agricultural credit may be defined-" as the amount of investible funds made available for the farm business and farmers' family needs." In other words, agricultural credit means money borrowed for farm investment as well as seasonal agricultural operations.

Agricultural credit is the money extended to the farmers to stimulate the productivity of the limited farm resources. It is not a mere loan or advance, it is an instrument to promote the well-being of the society. The role of farm credit in strengthening and developing both input and output markets in agriculture is crucial and significant. Providing appropriate finance for agricultural operations, therefore, like oiling agriculture to make its wheels more softly and smoothly. More so in the context of India, which isjam-packed with small farmers who have almost no resources and are in dire need of them.

In the past, agricultural credit was confined to a few needy farmers and a few elite moneylenders. During the drought conditions the farmers used to get food grains from rich farmers, on the conditions the same would be returned with interest after harvesting the next crop. Now days it has become a powerful instrument in bringing about agriculture and rural development. It has more significance in the case of marginal and small farmers as their savings are meager. They are in the vicious circle of poverty. Credit, if properly channelised, can break this vicious circle of poverty by providing them means to get productive assets or by providing employment opportunities. Hence, credit has a crucial role to play in raising the rural incomes.

The advent of modern technology has led to increased demand for inputs. As a result, adequate agricultural credit is required for agricultural development. As stated, provision of good credit facility to the farmers at the right time in the right place in the right proportion is a requisite for the transformation of agriculture. The non-availability or scarcity of credit retards agricultural progress in particular and rural development in general. An agriculturist has to make investment to raise the productivity of land. F.A. Nicholson's statement throws light on the importance of credit in agriculture. " The lesson of universal agrarian history from Rome to Scotland is that an essential of agriculture is credit. Neither the condition of the country nor the nature of land tenures, nor the position of agriculture affects the one great fact that agriculturists must borrow." Indian agriculture is still traditional subsistence and stagnant in nature, hence credit is needed to create the supporting infrastructure for adoption of new technology. Massive investment is needed to carry out major and minor irrigation projects, rural electrification, installation of fertilizers and chemical plants, implementation of agricultural promotion programmes and poverty alleviation programmes in the country.

As pointed, "credit makes it possible for farmers to take advantage of new machines, good seeds, fertilizers, livestock, labour, all of which enable the farmer to organise and operate his farm on more profitable basis. In order to adopt high yielding varieties, farmers need huge amount of loan, since their own savings are negligible. Expansion of credit institutions is a prerequisite for technological change, which will facilitate agricultural prosperity. The Green Revolution of Punjab is the best example to support this statement. Credit is a bad master if it is mismanaged. It is like the French proverb " credit supports the farmers' as the hangman's rope supports the hanged." It is considered as an economic ladder or elevator in uplifting the socio-economic conditions of the farmers. However, it serves as a useful means only when it is used judiciously for productive purposes otherwise it adversely affects on the repaying capacity of the farmers.

2.3 Classification/Types of Agricultural credit

2.3.1 Classification

Agricultural credit can be classified based on purpose, repayment period, security, generation of surplus funds, creditor and number of activities for which credit is provided.

- Purpose: Based on the purpose for which loan is granted, agricultural credit is categorized into:
  - Development credit or Investment Credit: This is provided for acquiring durable assets or for improving the existing assets. Under this, credit is extended for:
    - purchase of land and land reclamation.
    - purchase of farm machineries and implements
    - development of irrigation facilities
    - construction of farm structures
    - development of plantation and orchards
    - development of dairy, poultry, sheep/goat, fisheries, sericulture, etc.
  - Production credit: is given for crop, production: Here, the loan amount is used for purchasing inputs and for paying wages.
  - Marketing credit: It is essential to carry out the marketing functions and to get higher prices for the produce.
  - Consumption credit: It is the credit required by the farmer to meet his family expenses.
- Repayment Period: Based on the period for which the borrower requires credit, it is divided into:
  - Short-Term Credit: It is given to farmers for periods ranging from 6 to 18 months and is primarily meant to meet cultivation expenses viz., purchase of seed, fertilizer, pesticides and payment of wages to labourers. It serves as the working capital to operate the farm efficiently and is expected to be repaid at the time of harvesting / marketing of crops. It. should be repaid in one installment.
  - Medium-Term Credit: Repayment is for the period of 2 to 5 years, It is for the purchase of pump-sets, farm machineries and implements, bullocks, dairy animals and to carry out minor improvement in the farm. It can be repaid either in half yearly or annual installments.
  - Long-Term Credit: It is advanced for periods more than 5 years and extends even unto twenty five years against mortgage of immovable property for undertaking development works viz., sinking wells,

purchase of tractor, and making permanent improvements in the farm. It has to be repaid in half-yearly or annual installments.

- Security: Credit is provided to farmers based on the security offered by them.
- Farm Mortgage Credit: It is secured against mortgage of land.
- Collateral Credit or Chattel Credit: It is given against the security of livestock, crop or warehouse receipt.
- Personal Credit: It is given based on the character and repaying capacity of the person and not on any tangible assets. In general, LT credit is usually advanced against security of land while MT and ST loans are sanctioned against personal and. collateral security.

❖ Generation of Surplus Funds: Based on generation of surplus funds, credit can be classified as self-liquidating and non-self -liquidating credit.

- Self Liquidating Credit: In this case, loan amount gets absorbed in the production processing one year or production period and the additional income generated is sufficient to repay the entire loan amount.
- Non-Self Liquidating Credit: Here the resources acquired with the borrowed funds are not consumed in the production process during the project period. The investment is spread over a period of several years. The additional income generated in one year is not sufficient to repay the entire loan amount and hence the repayment is spread over to number of years.

❖ Creditor or Lender wise Credit: Credit can be classified from the point of view of creditor.

- Non - Institutional Agencies: They include money lenders, traders, commission agents, friends and relatives. This kind of loan is generally exploitative.
- Institutional Agencies: They include co-operative, commercial bank and regional rural bank.

❖ Number of Activities Served: Based on the number of activities for which amount the loan can be used, credit can be categorized as

- single purpose loan
- composite loan

2.3.2 Types

Considering the period and purpose of the credit requirement of the farmers of the country, agricultural credit in India can be classified into three major types:

(a) Short Term Credit:

The Indian farmers require credit to meet their short term needs viz., purchasing seeds, fertilisers, paying wages to hired workers etc. for a period of less than 15 months. Such loans are generally repaid after harvest.

(b) Medium Term Credit:

This type of credit includes credit requirement of farmers for medium period ranging between 15 months and 5 years and it is required for purchasing cattle, pumping sets, other agricultural implements etc. Medium term credits are normally larger in size than short term credit.

(c) Long Term Credit:

Farmers also require finance for a long period of more than 5 years just for the purpose of buying additional land or for making any permanent improvement on land like sinking of wells, reclamation of land, horticulture etc. Thus, the long term credit requires sufficient time for the repayment of such loan.

2.4 Need of Agricultural Finance Credit

2.4.1 Credit in Indian Agriculture

No sector in the economy can develop without the availability of timely and adequate finance. Agriculture is no exception to this. Since our agriculture is exposed to low returns and uncertainties due to its more reliance on nature. As most of the farmers are poor, it has become more imperative for them to borrow. It is difficult for our farmers to manage agricultural operations without resorting to borrowing. The following factors are responsible for the need of credit in Indian agriculture.

i. Low Income of farmers

The income of an agriculturist is very much lower than that of his counterpart in non-agricultural activity and it is a common experience of every farmer in the regional or national level. In a developed economy like U.S.A. or a developing economy like India, it is true that farmers receive a very much low income compared to others. As observed, income of the farmers is not comparable to the income of the clerical workers (low paid workers) in non-agricultural activities i.e the income of the rich in agriculture can be compared at the most with the income of the clerical workers in non-agricultural activity. This situation of comparatively low incomes has been characterised as one of the "farm problems." If this is the fate of an agriculturist in the U.S.A., one can easily imagine the deplorable plight of the proverbially poor Indian farmers.

The low income of the agriculturist naturally results in low savings, low investment, low productivity, low income and keep them in the vicious circle of poverty. It is clear, however, that the return to further investment in agriculture is normally low and low returns are likely to be an important limiting factor to increased savings and investment. But the farmers have to invest minimum amount for seasonal agricultural operations. In such situation, borrowing becomes inevitable. As pointed out to peculiar condition prevailing in Indian agriculture and calls it the "Paradox of growth." The per capita income registered an increase in all other activities except agriculture during 1970-71 to 1980-81.

The overall performance of Indian agriculture during the post reform period has been very poor. The study reveals that the annual compound growth rate of food grains production has declined from 6.36 percent in 1990-91 to 2.57 percent in 1995-96. The productivity of agriculture per acre as well as per worker has also witnessed a declining trend during the post-reform period compared with the same for the pre-reform period.

The growth rate of agricultural production has also been slower than that of our neighbours. According to World Bank Report, during 2008-2015, the annual agricultural growth rate was 3.24 per cent whereas those of China, Vietnam and Pakistan, were 5.18, 4.65 and 4.04 per cent respectively.

The Governments at the centre have been applying different standards for agriculturists and non-agriculturists, neglecting the former and protecting others. Hence, agriculture needs a fair deal. The provision of adequate and timely credit is essential for increasing the level of income of the farmers.

ii. Uncertain Income

Not only the income is very low in agriculture but also highly uncertain. So one cannot be sure of one's income and plan one's expenditure accordingly. Agricultural is susceptible to natural calamities and in countries like India, it is still depending on the vagaries of monsoon. Enormous risk in the form of uncertain income, bad weather, drought, flood, cyclone, pest attack, low prices, etc., have hampered agriculture. The return in agriculture is not certain. The operations of a farm enterprise are not as protected as in the case of commercial and industrial enterprises. Forecasting in non-agricultural enterprises is practiced, but the forecasting of farm income is a difficult task.

Agriculture suffers from various risks and uncertainties namely.

- the risks emanating from natural hazards and calamities such as earthquakes, floods, droughts, famines, hailstorms, tornados, damage of crops due to pests, insects and other diseases or loss of cattle due to accidents and infectious diseases, etc.
- the risk of loss of property by fire, thefts, and burglary, etc.
- loss due to abrupt and wide fluctuations in prices of farm products.
- risk arising out of the death or disability of the farmer. As observed,

"business of farming is the most precarious of all industries and it is very risky at the best."

A sectoral survey of incomes conducted in India during the period 1951-91 clearly reveals the uncertain nature of agricultural income visaversa other non-agricultural income. On the basis of survey, three important conclusions are drawn pertaining to agricultural income.

They are as follows.

- Negative growth is frequent in agriculture. During the period 1951-91, negative growth rate has been registered in fourteen different years in the GDP originated from agriculture, forestry, mining, and fisheries. But in the manufacturing, trade, transport, banking etc, negative growth rate has been registered only once.
- Wide changes in growth rate are common in agriculture. During the same period (1951-91), in 9 different years, nine times the change in the annual growth rate of GDP from agriculture was 10 percent or more in positive and negative direction but in industry, such huge changes in GDP were seen only 3 times that too in positive direction. In other activities, such wide changes were absent.
- In a short period of time, wide changes in opposite directions are possible in agriculture.

However, on the basis of these instances, one cannot infer that a very bad year is followed by a good year and visa-versa. Because, twice a negative growth rate continuously for two years and many times positive growth rate continuously for two or more years successively have been registered in GDP from agriculture, mining, forestry, fisheries etc. Thus, the survey supports the general notion that wide and unpredictable changes in income can occur in agriculture in any direction at any time.

Under these conditions, the minimum level of income which a farmer can earn, is not certain. The Agricultural Price Commission has been set up to fix the minimum prices for major agricultural crops, so that the farmers can be sure of at least the minimum price for their products. The study revealed that during the period of 20 years, the harvest prices were higher than the minimum support price in all the years for all the food grains except bajra in the year 1970-71. We know that the harvest prices are the lowest prices and if the minimum prices are invariably lower than the harvest prices, then it may be stated that minimum prices do not serve the desired purpose.

Under the conditions, farmers are very much in need of credit to diversify the crops, to develop allied activities, to set up household industries, to improve the conditions of farming, which may reduce the uncertainties in their income.

iii Increase in Expenditure

The cost of cultivation has been increasing due to the adoption of modem methods of cultivation, which is highly capital intensive. Modem inputs like improved seeds, fertilizers, pesticides, tractors, harvestors etc, are costly and to purchase such costly inputs, the farmers have to borrow. The adoption of new technology in agriculture would also influence the demand for credit substantially. Hence, for stimulating the tempo of agricultural production, it is necessary that farmers must be provided with adequate and timely credit. During mid sixties, the Green Revolution was introduced as a new package programme in the field of agriculture. It was a miracle in the field of agricultural production during 1968 in India through which we could produce the world's highest yield of rice and wheat. The production and productivity of land has increased tremendously. On the other hand the cost of cultivation also has been increased. Various studies supported this aspect. The study shows that in India variable input per acre is five times more in the case of new technology compared with that of traditional agriculture. As seen that, investment in durable capital can be seven times more in modern farms in comparison with traditional farms.

Agriculture is gradually becoming capital intensive and science based. New innovations like micro-irrigation system, selective agrochemicals, green house farming, tissue culture and many other techniques have made agriculture a very sophisticated profession. These new technologies are in a position to fully exploit the yield potentials of various crops. As the profession of agriculturist becomes more and more complex and technology-oriented, using the modern tools and techniques, it will be necessary for the agriculturist to borrow from institutional agencies for procuring these costly farm inputs.

iv. Typical Production Pattern

Agriculture being a seasonal activity, the farmer cannot start cultivating the land when he has funds but he has to spend for various agricultural activities during the particular season. Since agriculture is a biological activity, it is not possible to reduce the gestation period. However, hard one might work and one cannot get the output before the usual duration of time is over. Meanwhile, if one needs money, he has to think of other sources. In agriculture, there is a time gap between the investment and return. The gestation period of crops varies from months to years. To meet all the expenses till the receipt of income, credit is essential. The farmers' calculations do not always materialise. But the farmer who sows cannot be sure that he will reap. The precariousness of any line of farming renders it one of the most risky industries in the world, for no farmer can be certain that what he undertakes as a productive enterprise will turn out to be so. The risks of farming are incalculable. Though agriculture has such a production pattern, the persons who depend on it, can not have a similar expenditure pattern. They have to spend on a continuous basis on usual production and consumption.

The problems of finance and credit arise largely from a seasonal production cycle, which is super imposed on a non-seasonal consumption expenditure. Thus, provision for consumption and production inputs require either savings from the past harvest or credit borrowed against a future harvest.

v. Inability to raise funds through other sources

The farmers cannot raise funds for development by floating shares, as it is possible in other non-agricultural enterprises. Under the given conditions of poverty, the meager savings of the farmers cannot finance their requirements. Entering into a partnership with someone who has funds is not feasible in India. The major investments in agriculture are usually in the form of sunk capital which cannot be marketed so easily though a farmer wishes in times of emergency. He cannot sell or rent even a portion of his house or land in a profitable manner in the villages as there will not be much demand locally, a situation which is possible in urban areas. So whenever a farmer needs money, he has no option left but borrowing.

Under these conditions, credit is absolutely essential for the farmer. Even though his income is low and more than that he is compelled to increase his expenditure where it is not possible for him to raise funds in any other way. So certainty of expenditure and uncertainty of income make borrowing certain for the Indian farmers. In such a situation one has to agree with Sir Nicholson, "Credit is not only essential but also inevitable and as such it is not necessarily objectionable nor a sign of weakness".

vi. Stabilization of Income

The farmers must be able to increase their income. Therefore, they have to carry on allied activities like dairy, poultry, petty business in the villages etc, along with agriculture for which they need finance. They can stabilise the income with such allied activities. They can raise funds from different institutional agencies.

vii. Diversification and commercialisation of agriculture

Indian agriculture has been experiencing diversification and commercialization over the past several decades. The share of commercial crops like sugarcane, oilseeds, fibre crops, fruits, vegetables, flowers etc, in the gross cropped area has increased from 17.08 per cent during 1952-53 to 20.29 per cent during 1972-73 and 25.43 per cent during 1992-93. A distinct aspect of commercialization during the post-green revolution period was the introduction of new crops like sunflower, soyabean along with traditional food crops like groundnut, mustard etc. However, growing of food grains is still predominant mainly due to the farmer's orientation towards food security. The scope for diversification of food grains with vegetables, fruits and flowers continues to be immense in terms of farm profitability and employment of labour.

2.4.2 Need for credit to farmers

According to the World Bank, demand for food is estimated to increase by 70% by 2050 and at least USD$80 Billion will be needed in annual investments to meet this demand. A huge chunk of these billions is expected to come from the private sector and in particular the banking sector.

Some governments and farmers globally, are beginning to realize the pressing need to increase investment in agriculture. This is majorly because there are many unavoidable push factors at play including; population increase, climate change, advancement in agricultural technologies including innovations, increase in migrations, specific dietary needs and etc.

In developing countries, the banking sector apportions a much smaller share of its annual loan allocation to agriculture. This demonstrates a lack of trust and willingness by the private sector to invest in Agriculture while this is in part due to the sector's specific high financial risks including; production and price fluctuations, market risks, weather and climate change, among other major reasons.

Another major factor affecting financing of the Agricultural sector is the ineffective government policies, laws, strategies and other frameworks that promote and guide the financing component of the sector. Inclusive of these are policies and laws on land ownership and use, interest rate caps, loan management strategies and programs and concessional lending.

National budget allocations for Agriculture also make it difficult for the private sector to invest in it. This is so because they would have to trust a sector that sees a reasonably small amount of money to improve on components that would avoid major financial risks. Agricultural financing can take on different forms. It refers to financial services ranging from short term, medium and long-term loans, land leasing, crop and livestock insurance. All of these components of agri-financing cover aspects of agricultural processes from the planting to marketing phase of the sector value chain.

Credit is needed in every type of business and agriculture is no exception. The need for agriculture credit becomes more important when it moves from traditional agriculture to modern agriculture. Agricultural labour is often under-employed. Production suffers from weather risks. The capacity of farmers to save and invest is very low. The agricultural productivity is low due to low use of inputs. The farmers therefore, need credit to increase productivity and efficiency in agriculture. This need is increasing over the years with the rise in use of fertilizers, mechanisation and rise in prices. Some of the reasons why farmers need agricultural credit:

1. Purchase of new inputs

The farmers need finance for the purchase of new inputs which include seeds, fertilizers, pesticides, irrigation water etc. If the seed of high yielding varieties and other modern inputs are made available to the farmers they can increase productivity not only of land but also of labour.

2. Purchase of implements

Credit is required by the farmers for the purchase of tractors, threshers, harvesters, water pumping sets etc. The use of appropriate machinery in land will increase production by growing more than one crop on the same piece of land at the same time.

3. Better management of risk

Credit enables the farmers to better manage the risks of uncertainties of price, weather etc. They can borrow money during raining days and pay back the loans during peak years of crops.

4. Permanent improvement in land

Credit also helps the farmers to make permanent improvements in land like sinking of wells, land reclamation, horticulture, rotation of crops etc.

5. Better marketing of crops

If timely credit is available to the farmers, they will not sell the produce immediately after the harvest is over. At that time the prices of agricultural goods are low in the market. Credit enables the farmers to withhold the agricultural surplus an sell in the market when prices are high.

6. Facing crises

The credit is required by the farmers to face crisis. The crisis can be caused by failure of crop, draught of floods.

2.5 Role of Agricultural Finance Credit in Indian agriculture

2.5.1 Introduction

A large proportion of the population in India is rural based and depends on agriculture for a living. Enhanced and stable growth of the agriculture sector is important as it plays a vital role not only in generating purchasing power among the rural population by creating on-farm and off-farm employment opportunities but also through its contribution to price stability. In India, although the share of agriculture in real GDP has declined below one-fifth, it continues to be an important sector as it employs 52 per cent of the workforce. The growing adult population in India demand large and incessant rise in agricultural production. But per capita availability of food, particularly cereals and pulses, in recent years has fallen significantly. As a result, slackening growth of agriculture during last decade has been a major policy concern.

Three main factors that contribute to agricultural growth are increased use of agricultural inputs, technological change and technical efficiency. With savings being negligible among the small farmers, agricultural credit appears to be an essential input along with modern technology for higher productivity. An important aspect that has emerged in last three decades is that the credit is not only obtained by the small and marginal farmers for survival but also by the large farmers for enhancing their income. Hence, since independence, credit has been occupying an important place in the strategy for development of agriculture. The agricultural credit system of India consists of informal and formal sources of credit supply. The informal sources include friends, relatives, commission agents, traders, private moneylenders, etc. Three major channels for disbursement of formal credit include commercial banks, cooperatives and micro-finance institutions (MFI) covering the whole length and breadth of the country. The overall thrust of the current policy regime assumes that credit is a critical input that affects agricultural/ rural productivity and is important enough to establish causality with productivity. Therefore, impulses in the agricultural operations are sought through intervention in credit.

In order to improve the flow of credit to the agricultural sector, the Reserve Bank had advised public sector banks to prepare Special Agricultural Credit Plans in 1994-95. Under the SACP, the banks are required to fix self-set targets for achievement during the year (April- March). The targets are generally fixed by the banks about 20 to 25 per cent higher over the disbursements made in the previous year. With the introduction of SACP, the flow of credit to agricultural sector has increased significantly from Rs.8,255 crore in 1994-95 to Rs.1,22,443 crore in 2006-07 which were higher than the projection of Rs.1,18,160 crore. As against the target of Rs.1,52,133 crore for the financial year 2007-08, disbursements to agriculture by public sector banks under the plan were Rs.1,11,543 crore. As recommended by the Advisory Committee on Flow of Credit to Agriculture and Related Activities from the Banking System, the Mid-Term Review of Annual Policy of RBI for 2004-05 made the SACP mechanism applicable to private sector banks from the year 2005-06. Disbursements to agriculture by private sector banks under SACP during 2006-07 aggregated to Rs.44, 093 crore against the target of Rs.40,656 crore. As against the target of Rs.41,427 crore for the financial year 2007-08, disbursements to agriculture by private sector banks aggregated to Rs.45,905 crore (provisional).

From the Government side, with a view to doubling credit flow to agriculture within a period of three years and to provide some relief to farmers affected by natural calamities within the limits of financial prudence, the Union Finance Minister announced several measures on June 18, 2004. Accordingly, the Reserve Bank and NABARD issued necessary operational guidelines to banks. From the very beginning, the actual disbursements exceeded the targets for each of the last four years. As against the target of Rs.2,25,000 crore for 2007-08, all banks disbursed Rs.2,25,348 crore (provisional). During 2007-08, 7.29 million new farmers were financed by commercial banks and RRBs as against the target of 5 million farmers fixed by the Union Finance Minister for the year. The Finance Minister, in his Budget Speech for the year 2008-09, urged the banks to increase the level of credit to Rs.2,80,000 crore during the year 2008-09.

Such efforts have, however, not been transmitted to the growth in agriculture output. Since the mid-1990s, the growth of the agricultural sector has been low as well as volatile; the growth decelerated from an annual average of 4.7 per cent per annum during 1980s to 3.1 per cent during the 1990s and further to 2.2 per cent during the Tenth Plan period. Growth in agricultural production has decelerated during 2006-07 with the agriculture sector characterised by stagnation in output of major food grains. Per capita annual production of cereals declined from 192 kilogram (kg) during 1991-95 to 174 kg during 2004-07 and that of pulses from 15 kg to 12 kg over the same period. Per capita availability of food grains has, thus, fallen close to the levels prevailing during the 1970s.

Volatility in agricultural production has not only affected overall growth but also exerted persistence pressure on maintaining low and stable inflation. Demand-supply gaps were refl ected in higher domestic food prices in recent years. All these evidences apparently point to the fact that higher credit to agriculture is not translated into commensurate increase in agricultural output.

2.5.2 Approach

India has systematically pursued a supply leading approach to increase agricultural credit. The objectives have been to replace moneylenders, relieve farmers of indebtedness and to achieve higher levels of agricultural credit, investment and agricultural output. As per the study, the output and employment effect of expanded rural finance has been much smaller than in the nonfarm sector. The effect on crop output is not large, despite the fact that credit to agriculture has strongly increased fertilizer use and private investment in machines and livestock. High impact on inputs and modest impact on output clearly mean that the additional capital investment has been more important in substituting for agricultural labor than in increasing crop output.

Between bank nationalization in 1969 and the onset of financial liberalization in 1990 bank branches were opened in over 30,000 rural locations which had no prior presence of commercial banks. Alongside, the share of bank credit and savings which was accounted for by rural branches raised from 1.5 and 3 percent respectively to 15 percent each. This branch expansion was an integral part of India's social banking experiment which sought to improve the access of the rural poor to cheap formal credit. The estimates suggested that a one percent increase in the number of rural banked locations reduced rural poverty by roughly 0.4 percent and increased total output by 0.30 percent. The output effects are solely accounted for by increases in non-agricultural output – a finding which suggests that increased financial intermediation in rural India aided output and employment diversification out of agriculture.

The overall growth of agriculture and role of institutional credit with an overall supply of credit to agriculture serves as a percentage of total disbursal of credit that is going down, which should not be a cause for worry as share of formal credit as part of agricultural GDP is growing. This establishes that while credit is increasing, it has not really made an impact on value of output figures which points out the limitations of credit. Also, an attempt has been made to analyse the issues in agricultural credit in India which revealed that the credit delivery to the agriculture sector continues to be inadequate. It appeared that the banking system is still hesitant on various grounds to purvey credit to small and marginal farmers. It was suggested that concerted efforts were required to augment the flow of credit to agriculture, alongside exploring new innovations in product design and methods of delivery, through better use of technology and related processes. Facilitating credit through processors, input dealers, NGOs, etc., that were vertically integrated with the farmers, including through contract farming, for providing them critical inputs or processing their produce, could increase the credit flow to agriculture significantly.

In general, it is difficult to establish a causal relationship between agriculture credit and production due to the existence of critical endogeneity problem. However, it was concluded that increased supply and administered pricing of credit help in the increase in agricultural productivity and the well being of agriculturists as credit is a sub-component of the total investments made in agriculture. Borrowings could in fact be from multiple sources in the formal and informal space. Borrowing from formal sources is a part of this sub-component. With data being available largely from the formal sources of credit disbursal and indications that the formal credit as a proportion of total indebtedness is going down, it becomes much more difficult to establish the causality. He also stated that the diversity in cropping patterns, holding sizes, productivity, regional variations make it difficult to establish such a causality for agriculture or rural sector as a whole, even if one had data. Finally, he argued that mere increase in supply of credit is not going to address the problem of productivity, unless it is accompanied by investments in other support services. In the present study, we take a re-look at the problem by quantitatively assessing the impact of institutional credit expansion on agriculture.

2.6 Source of Agricultural Credit in India:

There are two broad sources of agricultural credit in India:

- Non-Institutional Sources
- Institutional Sources

Non-Institutional Sources

The non-institutional finance forms an important source of rural credit in India, constituting around 40 percent of total credit in India. The interest charged by the non-institutional lenders is usually very high. The land or other assets are kept as collateral. The important sources of non-institutional credit are as follows:

- Money-Lenders: Money-lending has been the widely prevalent profession in the rural areas. The money-lenders charge huge rate of interest and mortgage the property of the cultivators and in some cases even the peasants and members of his family are kept as collateral.
- Other Private Sources:
  - Traders, landlords and commission agents: The agents give credit on the hypothecation of crops which when harvested is used to repay loans.
  - Credit from relatives: These credits are generally used for meeting personal expenditure.

Institutional Sources

The general policy on agricultural credit has been one of progressive institutionalization aimed at providing timely and adequate credit to farmers for increasing agricultural production and productivity. Providing better access to institutional credit for the small and marginal farmers and other weaker sections to enable them to adopt modern technology and improved agricultural practices has been a major thrust of the policy. National Bank for Agriculture and Rural Development (NABARD) is an apex institution established in 1982 for rural credit in India. It doesn't directly finance farmers and other rural people. It grants assistance to them through the institutions described as follows:

1. Rural Co-Operative Credit Institutions:

Rural Credit cooperatives are the oldest and most extensive form of rural institutional financing in India. The major thrust of these cooperatives in the area of agricultural credit is the prevention of exploitation of the peasants by moneylenders. The rural credit cooperatives may be further divided into short-term credit cooperatives and long-term credit cooperatives. The short-term credit cooperatives provide short-term rural credit and are based on a three-tier structure as follows:

- Primary Agricultural Credit Societies (PACs): These are organized at the village level. These societies generally advance loans only for productive purposes. The main objective of a PACS is to raise capital for the purpose of giving loans and supporting the essential activities of the members such as supply of agricultural inputs at cheap price, improving irrigation on land owned by members, encourage various income-augmenting activities such as horticulture, animal husbandry, poultry etc. In India, around 99.5 percent of villages are covered by PACs.

- District Central Cooperative Banks: These cooperatives are organized at the district level. The PACS are affiliated to the District Central Co-operative Banks (DCCBs). DCCBs coordinate the activities of district central financing agencies, organize credit for PACs and carry out banking business.
- State Co-Operative Banks: The DCCBs are affiliated to State Co-operative Banks (SCBs), which coordinate the activities of DCCBs, organize provision of finance for credit worthy farmers, carry out banking business and act as leader of the Co-operatives in the States.

Long-term credit Cooperatives: These cooperatives meet long-term credit of the farmers and are organized at two levels:

- Primary Co-Operative Agriculture and Rural Development Banks: These banks operate at the village level as an independent unit.
- State Co-Operative Agriculture and Rural Development Banks: These banks operate at state level through their branches in different villages.

2. Commercial Banks:

Commercial Banks(CBs) provide rural credit by establishing their branches in the rural areas. The share of commercial banks in rural credit was very meager till 1969. The All India Rural Credit Review Committee (1969) recommended multi agency approach to the rural and especially agricultural credit. It suggested the increasing role of the CBs in providing agricultural credit. Further, under the Social Control Policy introduced in 1967 and subsequently the nationalization of 14 major CBs in 1969 (followed by another six banks in 1980), CBs have been given a special responsibility to set up their advances for agricultural and allied activities in the country. The major expansion of rural branches took place and CBs introduced Lead Bank scheme and district credit plans for rural areas. Banks were asked to lend 18 percent of their total advances to agriculture within the quota of 40 percent of priority sector lending. This expansion of rural credit remained till the late 1980s. However, during late 80's, CBs suffered huge losses due to waiving of agricultural loans by the government. The financial liberalization process with the adoption of Narasimham Committee report in 1993 has necessitated the banks to focus on profitability and adopt prudential norms. The proportion of bank credit to rural areas especially small borrowers has come down steadily.

3. Regional Rural Banks (RRBs):

RRBs are the specialised banks established under RRB Act, 1976 to cater to the needs of the rural poor. RRBs are set-up as rural-oriented commercial banks with the low cost profile of cooperatives but with the professional discipline and modern outlook of commercial banks. Between 1975 and 1987, 196 RRBs were established with over 14,000 branches. As a result of the amalgamation, the number of RRBs was reduced from 196 to 133 as on 31 March, 2006 and to 96 as on 30 April 2007.

RRBs covered 525 out of 605 districts as on 31 March 2006. After amalgamation, RRBs have become quite large covering most parts of the State. Increased coverage of districts by RRBs makes them an important segment of the Rural Financial Institutions (RFI). The branch network of RRBs in the rural area form around 43 per cent of the total rural branches of commercial banks. A large number of branches of RRBs were opened in the un-banked or under-banked areas providing services to the interior and far-flung areas of the country. RRBs primarily cover small and marginal farmers, landless laborers, rural artisans, small traders and other weaker sections of the rural community. However, even after so many years, the market share of RRBs in rural credit remained low and have suffered huge losses. In recent years Government has initiated reform process to improve the functioning of RRBs.

Efforts are made to increase the capital base and investible fund of these banks. The financial support is provided to improve training, technology development including computerization in these banks. The structural consolidation process has been initiated by amalgamating these banks. This has resulted into pooling of resources including experienced work force, common marketing efforts and thus better customer services. Further, RRBs have been able to derive the benefits of increased area of operations and enhanced credit exposure. These measures have provided remarkable improvements in the financial performance of RRBs. The number of RRBs incurring losses and the levels of non- performing assets has reduced dramatically.

4. Micro Finance Institutions (MFIs):

Banks offer concessional interest rates for the rural credit. However; small farmers are unable to access them because of borrower-unfriendly products and procedures, inflexibility and delay, and high transaction costs, both legitimate and illegal. Thus, Non-Government Organisations (NGOs) are providing alternative means to enhance access to credit by the poor since mid-70's. After pioneering efforts by organizations like SEWA, MYRADA, PRADAN and CDF, in 1992 the RBI and NABARD encouraged commercial banks to link up with NGOs to establish and finance self-help groups (SHGs) of the poor. The RBI has included financing of SHGs under priority sector lending. At present, there are three groups of SHGs viz. SHGs formed and financed by the banks (20 percent); SHGs formed by other formal agencies but financed by banks; SHGs financed by banks using NGOs and other agencies (8 percent).These institutions provide small loans to the poor at low interest rates without collateral.

The experience of micro-finance scheme in India suggests that

- ❖ It is the cost effective way of financing the rural poor;
- ❖ The repayment rate of SHGs is more than 95 percent due to peer pressure;
- ❖ It reduces transaction costs of borrowers as well as lenders;

- ❖ It inculcates the habit of thrift among members and provide timely credit.

2.7 Trends in Agricultural Credit

2.7.1 Introduction

Over time, spectacular progress has been achieved in terms of the scale and outreach of institutional framework for agricultural credit. Some of the major discernible trends are as follows:

1. Increasing Dependence on Institutional Credit

One of the major achievements in the post-independent India has been the widening of the spread of institutional machinery for credit and decline in the role of non-institutional sources. The share of institutional credit, which was little over 7 per cent in 1951, increased manifold to over 66 per cent in 1991, reflecting a remarkable decline in the share of non institutional credit from around 93 per cent to about 31 per cent during the same period. However, the latest NSSO Survey reveals that the share of non-institutional credit has taken a reverse swing which is a cause of concern.

Table: Relative Share of Borrowing of Cultivator Households from Different Sources

(Figures are in Per cent)

Sources Credit	1951	1961	1971	1981	1991	2002
Non-Institutional	**92.7**	**81.3**	**68.3**	**36.8**	**30.6**	**38.9**
of which Money Lenders	69.7	49.2	36.1	16.1	17.5	26.8
Institutional	**7.3**	**18.7**	**31.7**	**63.2**	**66.3**	**61.1**
of which Cooperatives Societies / Banks	3.3	2.6	22.0	29.8	23.6	30.2
Commercial Banks	0.9	0.6	2.4	28.8	35.2	26.3
Unspecified	-	-	-	-	**3.1**	-
Total	**100.0**	**100.0**	**100.0**	**100.0**	**100.0**	**100.0**

2. Trends of Overall Institutional Credit Flow

Over time, the flow of credit to agriculture and rural sector has expanded impressively (Table). The ground level credit flow had registered an increase from Rs. 1675 crore in 1975-96 to Rs.86891 crore in 2003-04 and further to Rs.203297 crore in 2005-06. This rate of growth was even higher than the growth rate of Gross Domestic Product (GDP) originating in agriculture.

Despite this growth, the credit needs of agriculture have not been met fully and overwhelming numbers of farm households have not been able to borrow from institutional sources. While short-term credit has remained the dominant component of total credit, its relative importance declined from 70.3 per cent in 1975-76 to 58.1 per cent in 2006-07.

Table: Institutional credit flow to agriculture sector (Rs. crore)

Agency	1975-76	1983-84	1993-94	2001-02	2002-03	2003-04	2004-05	2005-06	2006-07	2007-08*
Cooperatives	1186	2938	10117	23524	23636	26,875	31231	39404	42480	33070
Regional Rural Banks	2	263	977	4854	6070	7581	12404	15223	20435	15925
Scheduled Commercial Banks	405	1885	5400	33587	39774	52441	81481	125477	140382	88765
Other Agencies	82	185	0	80	80	84	193	382	NA	NA
Total Credit	1675	5244	16494	62045	69560	86981	125309	180486	203297	137760

3. Agency-wise Credit Flow

The analysis of agency wise credit flow indicates that the cooperative banks were the major source of agriculture credit in 1975-76 constituting around 71 percent of the total ground level credit flow followed by commercial banks at 24.2 percent and regional rural banks at 4.9 percent . Though cooperative banks had dominated agriculture credit supply till the early reform period, commercial banks and RRBs recorded impressive growth rates. As a result, in 2006-07, the share of cooperative banks in the total institutional credit flow receded to 20.1 percent and that of commercial banks advanced to 69.1 percent. Although the quantum of disbursement from cooperative banks increased, it could not keep pace with commercial banks in enhancing credit flow due to several reasons including its poor financial health, dual control, lack of internal controls and corporate governance norms and excessive dependence on other financial institutions.

Although the share of cooperative credit is now much lower than that of commercial banks, the reach of cooperative credit societies is much wider. Cooperative credit societies have more than twice the number of rural outlets and four times more accounts than those of scheduled commercial banks and RRBs put together. Low recovery rates and mounting overdues have clogged the process of recycling of credit by cooperatives, impaired their ability to avail of refinance facilities from (NABARD), increased transaction costs and more importantly, have deprived potential borrowers of the opportunity to avail of credit facilities from the cooperatives. As a result, cooperatives have been losing their capacity to meet the growing credit needs of agriculture.

4. Size-wise Credit Flow

Despite impressive growth in direct credit to farmers from the scheduled commercial banks between 1991-92 and 2003-04, contrary to expectation, credit disbursement to small and marginal farmers has not been encouraging. Though the number of accounts increased for small farmers yet the credit flow favoured the richer farmers.

5. Region-wise Credit Flow

While analyzing the pattern of credit flow, it is observed that the proportions of bank deposits and credit shares have moved in favour of the South, West and North regions. While the share of loans in the total disbursement of credit for agriculture and allied activities were the maximum for the South region, it was the minimum for North-east region.

2.7.2 Activities

Government of India & Reserve Bank of India: In order to increase credit flow to the agriculture sector, the policy of doubling of agricultural credit in three years was introduced in 2004-05. In order to expand the outreach of the banking services, banks made available basic banking 'no-frills' account with low or nil minimum balances as well as low or no charges in 2005. The regional rural banks were also specifically advised to allow limited overdraft facilities in 'no-frills' accounts without any collateral or linkage to any purpose.

National Agricultural Insurance Scheme (NAIS): NAIS is implemented since Rabi 1999-2000 season with the objectives to provide insurance coverage and financial support to the farmers in the event of failure of any of the notified crops as a result of natural calamities, pests and diseases and to encourage the farmers to adopt progressive farming practices, high value in-puts and higher technology in agriculture and to help stabilize farm incomes particularly in disaster years.

Government of India & NABARD

i. Rural Infrastructure Development Fund: RIDF was established in 96-1995 with a corpus of Rs. 2,000crores with the major objective of providing funds to state governments and state owned corporations to develop rural infrastructure such as rural roads, rural bridges, irrigation works, soil conservation, flood protection, drinking water, infrastructure for rural education etc. The total corpus of RIDF till 08-2007 (RIDF-I to RIDF-XIII) amounted to Rs.000 ,72 crores with 09-2008 budget further adding the amount of RIDF XIV of Rs. 14,000 crores to this corpus. The total sanctions and disbursements as on 30 June 2007 aggregated Rs. 61312.27 crores and Rs.38581.82 crores respectively.

ii. Micro Finance Innovations: The credit accessibility for the poor from conventional banking is limited due to lack of collaterals and information. Micro finance has emerged as an alternative financial vehicle that provides micro credit or small loans granted to the poor without any collateral. These loans are provided through micro finance institutions (MFIs). NABARD plays a key role in developing the MFIs by providing them refinance facility at low interest rates.

iii. Kisan Credit Card Schemes: The kisan credit cards (KCC) scheme was introduced in 99-1998 to facilitate short-term credit to farmers. Each farmer is given with a kisan credit card and a pass book for providing revolving cash credit facilities. NABARD provide refinance facility to commercial banks and cooperatives to provide credit under this scheme.

iv. Refinance under Swarnajayanti Gram Swarozgar Yojna (SGSY): NABARD provides refinance facility to institutions that support SGSY.

v. Co-operative Development Fund: NABARD has set up the cooperative development fund in 1993 with objective of strengthening the co-operative credit institutions in the areas of resource mobilization, recovery position etc. the assistance is provided to cooperatives by way of soft loans or grants.

2.7.4 Weaknesses in Credit Structure

1. Overemphasis of Monetary Credit: The rural credit institutions have given overemphasis on the financial assistance to the cultivators. While the finance is very important factor but it should be complemented with the extension of services in form of guidance, expertise and counseling on agricultural issues.

2. Multiplicity of Institutions: The rural credit structure is based on multi agency credit system whereby there exist numerous organizations providing similar kind of financial services. There is a lack of coordination in the system and the commercial viability is adversely affected in this scenario.

3. Lack of Motivation: In order to fill the gap that occurred due to the failure of rural cooperative societies Government gave increasing role to the commercial banks. However, commercial banks lack the desired skills and expertise in the agro-credit. The banks have enough financial resources but the service consultancy is not available. Thus, there is a failure to provide complete package of assistance to the farmers. Further, financial sector reforms have put pressure on banks to improve their financial position and so these banks are now concentrating on selected clientele of large borrowers

4. Financial Exclusion: Despite of a large network of the institutional credit system, it has not been able to adequately penetrate the informal rural financial markets and the non-institutional sources continue to play a dominant role in purveying the credit needs of the people residing in rural areas. The results of the All-India Debt and Investment Survey also indicate that the share of the non-institutional sources, in the total credit of the cultivator households, had increased from 30.6 percent in 1991 to 38.9 percent in 2002.

5. High Interest Rates: The rate of interest charged by rural financial institutions (RFIs) from farmers continues to be considerably higher than those charged by financial institutions from urban consumers. The owners of small or marginal farms, which are non-viable or viable at the margin, and self-employed in the informal sector, cannot afford to bear the level of interest charged by RFIs.

6. Procedural Delays: There is a problem of considerable delays in processing of loan applications and collaterals. Thus farmers shy away from institutional financing and increase their dependency upon non-institutional sources.

7. Poor Recoveries: Banks are shying away from rural financing mainly because of poor recoveries which is inflicting the system. It is ironical that the recoveries position is adverse amongst rich farmers than amongst the small farmers. The political decisions of waiving off loans are further putting pressures on the financial system.

2.7.5 Steps for Improvement in Institutional Credit System

1. Financial Discipline to Improve Recovery: A national consensus among political parties should be evolved for not politicizing the RFIs and resist from announcement of loan or interest waiver schemes and giving calls for not repaying the institutional loans. However, given the risk involved in the agriculture credit the recovery system should be flexible and humane.

2. Revamping the Cooperative Credit Structure: The Cooperative Credit Structure should be strengthened to make use of its wider reach. These have to be recapitalised so as to provide funds for improving their financial positions. There is a need of capacity building, human resource development, institutional restructuring to ensure democratic functioning, and improving the regulatory regime to empower the Reserve Bank of India (RBI) to enforce prudent financial management.

3. Better Physical, Social and Economic Infrastructure: The long term policy framework needs to be designed to improve infrastructure facilities so as to boost rural economic growth. This requires **increased public expenditure**on social infrastructure (like education, availability of drinking water, health facilities), physical infrastructure (like roads, power) and economic infrastructure like (irrigation, modern agricultural techniques). These measures would help to improve the debt paying capacity of rural poor and provide greater opportunities to RFIs.

4. Financial cum Consultancy Approach: RFIs needs to provide extension services like consultancy about seeds, availability and use of modern inputs, marketing strategies etc to the cultivators so that a holistic package of assistance can be provided to them.

5. Group Approach to Lending: The lending to homogenous farmer's groups needs to be organized to improve credit delivery. This would help to improve recovery because of peer pressure. Further, group lending tends to be cost-effective. Involving NGOs or rural educated youths in organizing farmers or rural families in groups, scrutinizing applications, disbursement of loan and effecting recoveries would help RFIs in reducing lending costs.

6. Autonomy to RRBs: RRBs should be given more autonomy and flexibility in planning and lending policies, so that their comparative advantage in rural lending is restored.

7. Greater involvement of Micro Finance Organizations: The banks need to involve micro-finance agencies like SHGs, NGOs etc. and other grass root level financial intermediaries who have better understanding of the credit needs and recovery situations.

8. Technological Up Gradation: Technological improvements like computerization can be critical in building up a reliable credit information system and database on customers, reducing transaction costs and facilitating better pricing of risk, improving the efficiency of the financial system, and thereby increasing the access of un-banked rural people in an efficient manner.

9. Information Dissemination to Rural Poor: Credit counseling, awareness and financial education regarding the benefits of institutional financing are important for effective expansion of financial services in rural areas. To do this, banks may utilize the services of non-governmental organisations, village youth clubs, village panchayats, farmer clubs and self-help groups into confidence.

2.8 Problems of Agricultural Credit

2.8.1 Conditions of Agricultural Credit Institutions

Credit is inevitable for agriculture. But provision of credit in a satisfactory way is not easy for the credit institutions in the present context. The major factor responsible for such a state of affairs seems to be the widening gap between what is expected and what is possible (reality) in the case of agricultural credit institutions.

To begin with the nature of agriculture itself is such that it poses many problems both to the farmers and the financing agencies. The meaning of agricultural credit has been widened to encompass several aspects. The efforts to revitalise the agricultural credit agencies to cope with the situation are not keeping pace with the increasing load they are supposed to carry. The agricultural credit scenario is not promising as it is expected. They are supposed to compete in connection with the agricultural credit in the changed economic situation.

The role of credit institutions in the field of agriculture is not confined to just lending but much more than that. It is the responsibility of the credit system to make available all the necessary facilities at the door-step of the farmers, train them to achieve rapid agricultural development. The objectives of good credit system are huge deposit mobilisation, high deposit credit ratio, easy disbursal of the amount at right time for right purpose to right persons, viability, maximum recovery, reasonable rate of interest, minimum cost etc. No credit agency has got all the above features. The functioning of agricultural credit system is too complicated. If all the above features are incorporated in the agricultural credit system, undoubtedly the system will emerge as an ideal one. It is the pertinent issue of credit to become ideal under the present conditions of agriculture.

1.Inadequate Facilities

Agricultural credit institutions need a conducive atmosphere to function at the expected level, comprising favourable socio-economic set up which depends to a considerable extent on the government policy. In other words, for the smooth functioning of the agricultural credit institutions, improvements in basic infrastructure are needed. It is the common experience of all the credit institutions that the farmers as a group prosper, along with it the banks easily flourish and vice-versa. An increase in the farmers' income results in more deposits, better recovery, huge borrowings etc, and thereby credit agencies grow faster. For the prosperity of the farmers, infrastructural facilities have to be improved which includes:

i) Marketing

The improvements in marketing are essential for the prosperity of the farmers. The marketing infrastructure is inadequate and existing conditions in the markets lead to the exploitation of the innocent and ignorant farmers. The fair prices may not be provided to their products.

Major share of the benefits of the agricultural produce goes to the middlemen. Many studies also have corroborated this aspect. Therefore, the farmers have to be suffered in the present conditions. It adversely affects the functioning of the rural credit institutions.

ii) Technological Research

The technological innovations are conducive to the increase in the productivity and production of agriculture. As stated, there are ample evidences to say that new technology exerts a positive influence on productivity even when the effects of weather are adverse. Agriculture is a location of specific activity. So the technology invented has to be tailored to suit the climatic conditions of a given area to better results. To achieve this, research must be conducted at the regional levels for various crops that are grown in specific area.

However, the present condition is not encouraging in the promotion of technological research in the field of agriculture. The wide spread use of new technology involving intensive application of modem inputs has given rise to increasing demand for credit from agricultural sector. The farm technology without adequate credit is just as a high cost dam without field channel for irrigation. Hence the availability of credit is necessary to sustain the technological progress in agriculture.

iii) Insurance

There are several risks and uncertainties associated with agriculture. The Government has introduced the Crop Insurance Scheme for some crops for which the farmers can get insurance facilities. The rebate in the premium is also given to the small and marginal farmers. But most of the farmers do not have the knowledge of such schemes. It is not so effective. The procedure of such schemes is very complicated.

The farmers should produce a certificate from the revenue department stating that in two out of three successive years, the crop grown in a given area was less than 50 per cent of the average yield in the area. On the basis of such a certificate, the farmers as a group can avail themselves of the insurance benefit. Obviously, the method of availing of the insurance benefit is not veiy simple, which prohibits many deserving farmers from benefiting the scheme. It is necessary to give such benefits to the farmers when the prices of the products are very low. In countries like USA, even life insurance is a part of farm credit. It was suggested that group insurance scheme for all the borrowing farmers, who borrow from a particular credit institution which works out cheaper by handling the insurance on a wholesale basis. In India the facility of combining agricultural credit with different types of insurance schemes is not adequate.

2. Low Investment in Farm Sector by Institutional Agencies

The commercial banks' lending to farm sector has declined from 16.3 percent of net bank credit in 1996-97 to 15.7 percent in 1997-98 as against the target of 18 percent though they have a veiy comfortable resource position and good network of branches in rural and semi-urban areas. The RRBs have also reduced their lending to farm sector in particular and rural area in general. Their credit deposit ratio declined from 50.3 percent in 1997-98 to 43.3 percent in 1998-99. This gives an impression that both institutions are inclined to withdraw from this field. This may be more pronounced when government share holding would be reduced to 33 percent. It is therefore, necessary to have a look at bank's specific problems in order to appreciate the need for improving the financial health of the banks.

3.High Non-Performing Assets (NPAs) of Co-operatives

Resource base of the co-operative credit institutions has been poor necessitating dependence on their higher tiers and refinancing agencies.

Overdue amount was as high as Rs.9925 crores on June 1998 in the short-term wing of the agricultural credit structure. In case of long-term wing the overdues were of the order of Rs.2962 crores in 1998. SCBs and DCCBs have built-up NPAs to the extent of Rs.2748 crores and Rs.6573 crores respectively in June 1999, whereas state and primary level banks in the long term wing have NPAs of Rs. 1977 crore and Rs.637 crore respectively.

Lack of required follow-up and supervision as well as sponsored lending programme and loan waiver and write off schemes have been much more a cause of concern for mounting overdues and thus creating NPAs.

4.Security Against Loans

The institutional agencies try to distribute their loanable funds in such a way as to minimise risk. Before making an advance the banker satisfies himself whether the borrower has the security to be pledged to ensure the repayment of credit, whether he has the capacity to derive the benefits from the use of funds and whether he is willing to repay.

Since adequate information about the borrowers' personal credibility is not easily available, the lending agencies insist more on tangible assets as security against loans. Under these circumstances, credit tends to be rationed not according to the needs but according to the existing economic status of the borrower. For this reason, it is sometimes believed that the small and marginal farmers have to depend upon non-institutional agencies of credit whose lending operations are convenient to the borrowers. Since millions of small and marginal farmers contribute a significant portion to the total agricultural output in our country, provision of adequate and timely credit for them should definitely be considered as an integral part of the agricultural credit policy.

5.Diversion of Production Credit to Consumption

The diversion of credit for purposes other than for which they are borrowed is a phenomenon in agricultural financing. It is often alleged that because the small fanners have higher propensity to consume, they divert production credit to consumption, thus constituting a high risk in repayment of credit. Even if they use credit for production, because of their higher marginal propensity to consume, the increased income may be used for consumption. If the credit is used for productive purposes, the credit agencies may not be suffered in case of recovery of dues. The lender and borrower have to exercise great care in seeing that the accommodation does not lead to wasteful expenditures

6. Ineffective Utilisation of Credit

The problem of agricultural credit in our country is not merely one of inadequate supply of funds rather it is aggravated by its ineffective utilisation. Credit serves useful purpose only when it is used for productive purposes to generate a surplus after paying interest and capital. Its diversion for unproductive purpose would affect the repaying capacity of the borrower, which may further cause the problem of overdues and defaults.

7. Lack of Effective Inspection

Agricultural credit needs to be reviewed in a dynamic context. Credit, can be dynamic only if it generates additional income for the farmer. Being production oriented the proper utilisation of credit must result in additional output and income, which can be used to repay old loans and make further investments for the improvement of farm organisation. The proper end-use of credit should therefore, be the keynote of institutional finance to agriculture, for it has a direct bearing on the quality of lending and eventually the recovery performance neither of which any credit institution can afford to underestimate the emphasis has to be on quality of inspection rather than on the periodicity of inspections. The lending institutions should evolve well-structured inspection procedures, which should aim at meaningful and productive follow-up advances.

8.Typical Situation for Co-operative Banks

The co-operative banks have to work under the control of two Acts, one being the Banking Regulation Act 1949 and the other being the State Co-operative Act. If they are expected to compete with RRBs and commercial banks, then they have to be given a level playing field.

There does not appear any difference between co-operative bank and any other bank as far as business is concerned. They have to work under the control and guidance of NABARD. The committee has also stressed the problem of duality of control as the co-operative banks are regulated by the state government concerned, RBI and NABARD. The poor performance of co-operatives has been largely attributed to the duality of control.

9.Low Level of Resource-base

Another major problem of the co-operative banks has been the low level of resource-base. Certain policies on refinancing by higher apex agencies including NABARD have been responsible for this to a certain extent. These banks have not made sincere efforts to mobilise deposits. In fact the management continued to treat resource mobilisation as a low priority item as long as refinance as cheap source of funds is available to them. The cheap funds by way of refinance have been responsible for killing the initiative of mobilisation of deposits by co-operative banks.

The co-operative system is so much dependent on financing by higher agencies. If a co-operative spirit is to be infused in the co-operative banks, refinance should be made at the market rates. This may increase the losses of the banks initially, but would be a rigid strategy in the long run. In south Korea, the Apex Bank provides refinance to the Cooperative Banks at the co-operative market rates. This compelled the banks to concentrate more on mobilising their own cheaper resources from market.

10. Defective Loaning Policy

Regarding the short-term structure, a labourious policy is laid down to the PACS for preparing the credit limit statements and inspite of observing this formality the members are not assured of the loan in time due to various reasons. Various disciplines pertaining to credit limits minimsing for seasonal agricultural credit limits are imposed by NABARD, these conditions act as constraints on full utilisation of credit sanctioned by NABARD. The realisation of these conditions needs to be considered. The credit limits are to be prepared and submitted to the financing institution well in advance. But the DCCBs are not getting these applications even up to June. Some times the farmers may not get credit at right time. If they get later, there is the possibility of misusing it for non-farm activities. The PACS are advancing a portion of the farmers' requirements. Therefore, they are forced to go to either moneylenders or some other sources. Some times the amount sanctioned for 'B' component may not be useful to raise the crop.

11.Mounting Overdues

A highly distressing fact of co-operative credit is heavy overdues of co-operative credit institutions, estimated to Rs. 10,000 crore. The failure of crops is a common sight every year due to the natural calamities.

The repaying capacity of the farmers has been crippled. The Government has come to the rescue of the drought-stricken farmers by waiving the dues of interest during the drought years. But the NABARD is insisting upon a minimum recovery of 40 percent of the demand of the co-operative year to gain eligibility for the sanction of credit limits fixed by the NABARD as a matter of financial discipline. The waiving of loans by the government in the past for political reasons was the major problem and it punished the non-defaulters rather than defaulters Accounting practices of DCCBs which appropriate all recoveries first towards interest and then towards principal also tend to compound the problem. The DCCBs have no prospects of recovering the outstanding amount from PACS which are not recovered by the loans outstanding against ultimate borrowers.

Consequently, the recovery of DCCBs deteriorated and irrecoverable, as the outstanding amount increased. Over a period of time, the DCCBs have become ineligible for NABARD refinancing, as they would not meet the stipulated recovery percentage. It badly affects the credit flow to the PACS.

Thus, the institutional credit agencies are facing many problems in connection with the agricultural credit. In lieu of liberalisation and privatisation, there is an urgent need to redefine the government's role in developing and strengthening credit co-operative institutions.

2.8.2 Removing limitations

In order to remove limitations and problems of agricultural credit in India the following measures may be suggested:

- To monitor the taccavi loan offered by the Government in a serious manner.
- Co-operative credit societies should be organised to make it efficient and purposeful for delivering the best in terms of rural credit. Moreover, these societies may be transformed into a multi-purpose society with sufficient funding capacity.
- Middlemen existing between credit agencies and borrowers should be eliminated.
- Reserve Bank of India should arrange sufficient fund so that long term loans can be advanced to the farmers.
- Power and activities of the Mahajans and moneylenders should be checked so as to declare an end to the exploitation of farmers.
- The Government should introduce the credit guarantee scheme so as to provide guarantee on behalf of the farmers for getting loans.
- The banks should adopt procedural simplification for credit delivery through rationalisation of its working pattern.
- The Government should issue Kisan credit cards to the farmers to draw cash for their production needs on the basis of the model scheme prepared by NABARD.
- In order to check the fraud practices adopted by the farmer, for getting loans from different agencies by showing same tangible security, a credit card should be issued against each farmer which will show the details about the loans taken by them from different agencies.
- Credit should also monitor over the actual utilisation of loans by developing an effective supervisory mechanism.

3

Agricultural Credit Analysis

3.1 Introduction

Agriculture has its own inherit risks and unique characteristics. As a result, agricultural credit is different than normal consumer credit. It provides following to aid farmers and ranchers in understanding and maintaining their credit. Due to the nature of agriculture, the following guidelines are for Ag borrowers, and due to the complexity of agriculture some of the suggestions may not apply to varied situation. Risk is any lenders primary concern where lenders determine what the chances are that you will have late payments, and/or they will be required to liquidate collateral to collect payment, and/or they will lose some portion of the loan. All quality lenders want to make loans, but only those loans that provide assurance of timely repayment.

Farming operations, like any business, need money to operate. It is the essential element to running daily tasks, purchasing supplies, paying employees, and buying equipment. The typical farming operation runs into the problem of crop cycles and volatile income; a crop is grown for months without producing any revenue for the grower. Then once the harvest is complete, the grower begins earning revenue. Price volatility within the farming industry always provides growers with a level of uncertainty about how much cash will be earned each year by their crop. External factors such as weather and natural disasters can play a large role in the amount of crop that can be sold by farmers. All of these factors attribute to one of the largest obstacles farmers continuously face: uneven, unpredictable, and volatile cash flows. Cash flow problems can stymie production on a farm, especially on a capital intensive farming operation.

To avoid cash flow problems, borrowing money is essential to keeping the operation running smoothly. Many institutions are available for such a need: commercial banks, credit unions, and the Farm Credit System (FCS). Commercial banks, such as Wells Fargo and Bank of America, tend to have small agriculture divisions to provide lending for farmers. Agricultural lending in these institutions is a small portion of the lending portfolio, but offers diversification. Members of the FCS, such as American AgCredit and Yosemite Farm Credit, are enterprises that were established in 1916 by the federal government as a way to ensure there would always be money available for agriculture in America. These FCS institutions are set up as cooperatives. Therefore, the "owners" of the company are the farmers, or customers, who borrow money from the FCS institution.

When lending money to farming operations lenders will typically ask applicants to submit financial statements from the previous three years. The submitted financial statements come in many forms: balance sheets, income statements, cash flow statements, and possibly federal tax returns. The submitted financial statements are analyzed by the loan officer servicing the loan using liquidity ratios, probability of default scores, monthly and yearly cash flows, repayment capacity ratios, and collateral appraisal. The analysis of these tools is both subjective and objective. The loan officer's personal experience and expertise generally plays a large role in the approval/rejection decision of a loan request.

Agricultural loans come in two forms: operating loans and real estate loans. Operating loans are necessary for farmers because it provides them with a guaranteed amount of money for the entire year they will need to keep the farm operating properly. Simply saving money from prior years' profits is not sufficient because of the volatility of farm revenue from year-to-year. Any number of things can affect the outcome of a year's harvest: weather, increased operating costs, decreased prices, labor shortages, etc. These variables change every year making it very difficult to simply save money as a farmer. If a farmer has three consecutive years with a low yield due to poor weather conditions, there would be little or no profit to use for the following year's operation. Hence, operating loans are extremely common because they help mitigate the risks involved with farming.

3.2 Intent, lending and risk

3.2.1 Intent of Source Information

The analysis of an agricultural loan proposal is a process known throughout the credit industry, but is typically not common knowledge. This process does not use statistical testing or conventional data collection. Therefore, the content of my literature review will be explaining the nature and current state of the agricultural lending industry, general practices within the agricultural lending industry, and predictions of what the agricultural industry will be facing in the near future. I will be using the information or results presented in cited sources as a guideline of how to perform a credit analysis of an agricultural loan.

3.2.2 Background on Lending Industry

Agricultural lending is a crucial component in agricultural production because it helps mitigate farming risks and provides guaranteed funding for the yearly operation. Like any other business, farmers need money to acquire assets, maintain cash flows, operate consistently, and to expand. Some farming operations simply need loans to survive, while others use agricultural loans to thrive and expand. Loans are essential in the agriculture production industry mostly because the annual cycles involved with agricultural production. Most agricultural operations produce a product for only a portion of the year, which leads to irregular revenue streams, reliance on seasonal weather, and sometimes seasonal demand.

Risk plays a very important role in agricultural production because there are so many variable factors that go into agricultural production. Weather poses the biggest risk for farmers because it is uncontrollable, can be unpredictable, and very destructive. There is also a variance in input costs from one year to the next. Farming operations do not have cheap start-up costs either. Depending on the farming operation, the start-up just to begin producing a product can far surpass the savings of a single farmer. Therefore, loans in the agricultural production industry play a critical role.

Agricultural loans are broken into two categories: real estate loans and operating loans. Third-quarter reports in 2012 will show that lending to food producers is on an upswing. Bankers say the increased demand is derived from the increased beef consumption in the U.S.. This increase in consumption means beef producers need to sell more beef, resulting in a necessity for more up-front cash in order to raise the beef before turning a profit. This also means the beef producers need more feed, which is primarily comprised of corn. Therefore, corn producers need to increase production, resulting in more up-front cash as well. Beef producers are not the only producers who use corn in their feed. Producers of all kinds are seeing an increase in feed prices, which are affecting overall profits and ability to stay afloat.

A process called syndication has been shown in the French lending market to improve lenders' interactions, reciprocity, trust, experience, and reputation, which results in reduced information costs. Syndication is the collaboration of multiple lenders to provide a borrower with capital. This process is used when a borrower's needs exceed the resources of just one lender. Typically, a borrower's needs exceed resources when the loan amount is very large or the borrower's risk is too high for one lender to take on. Therefore, multiple lenders take their share in the risk and the profit. This collaboration helps improve efficiencies by cutting down administrative costs for a loan that will yield a large return. In the French lending market, reduced information costs resulted in a reduction of borrowing costs. This idea of syndication should be considered in the U.S. agricultural lending market so increasing demand can be taken advantage of by reducing all possible borrowing costs.

3.3 Factors determine credit- 4 C's of Credit

There are four factors that lenders look at to determine credit;

- character
- capital
- capacity
- collateral

1st C of Credit -Character

One of the most overlooked aspects in obtaining credit approval is the section of the credit application 4 C's of Credit that deals with character. Character is vital in the review of an applicant. The lender needs to be comfortable with who they are loaning their money to. Management is a reflection of character. Does the history of operation (including consumer credit) indicate good planning and discipline and if so, then many lenders will conclude about their excellent character. When lenders see historical indications of problems, then management ability about operator is questioned. Many problems are unavoidable and explainable, such as extended drought or hail damage, but repeated non-payment of debts is normally an indication of poor planning and/or lack of discipline regarding credit availability. Poor management equals poor character.

Reviewing a credit application with following key points in mind will go a long way toward getting a loan application approved which includes:

- A completely filled out application gives the lender a good indication of the applicant's stability.
- Make sure the applicant has a valid driver's license, included on the application. This may sound basic to the loan process, but many times it is overlooked. A driver's license helps a lender locate an applicant should issues arise.
- Always look for five years of residency.
- Is your applicant a home owner or a tenant paying rent? In the case of renters, what is the rental amount and to whom is it paid every month? A landlord name and phone number should be provided for verification.
- Regarding employment, the standard is five years. If the applicant is new to the job market; e.g. a new graduate, "student" should be listed as the previous occupation, including school attended and for how many years. If the customer changed jobs, but not job fields, a similar job description for each period of employment should be listed. Try to determine why they changed one job for the same work somewhere else. Account for any gaps in employment

- Finally, references are very important. The applicant should provide the correct number of personal references. Look for the lenders preference for family or non-family references, or a combination of both. Use local references and make sure addresses and phone numbers are accurate.

2nd C of Credit - Capital

Capital is examined from two perspectives, working capital and equity capital.

Working capital is determined by examining current assets and current liabilities; it tells lenders what your immediate repayment ability looks like. You need to have sufficient current assets to pay all your current liabilities with some reasonable margin of safety. It is impossible to utilize short-term loans (line of credit) to improve your working capital. True, a line of credit will provide necessary availability of cash, but remember that lines of credit are normally due at the end of the production cycle. Consequently, all you have done is increased both current assets and current liabilities.

Equity capital is your net worth; it reflects your historical management of assets. Equity capital is also the most critical item needed to weather financial storms. Good AG lenders know that bad years will occur, and the question is not if, but when they will occur. Your equity allows your lender the opportunity to see you through the problem. Without equity capital, when that bad year comes along, your lender will have no choice but to liquidate collateral. Net worth or equity capital is also examined from the perspective of what is earned and what occurred from appreciation of real property.

3rd C of Credit - Capacity

Capacity is ability to service the debt, and is also known as net available cash flow. Some borrowers are shocked to find they are denied because they don't have sufficient cash flow to service their debt, and can't understand why, because they have been making all their payments.

Historical repayment without sustainable cash flow can occur from two different sources; increased borrowing and/or liquidation of assets. It is necessary when during bad year to increase the borrowing as it may be wise to liquidate some assets from time to time. However, in order to sustain operations and provide timely repayment, then overall operation should generate sufficient excess cash (over and above expenses) to adequately support all debt payments. The source of this cash needs to be from the operation, not from borrowings or the sale of assets.

4th C of Credit - Collateral

Collateral is what serves as security to lender. By providing collateral to lender, if this doesn't work and he/she fails to make payments, in such case, the ownership of the item gets lost. Collateral provides for the shared risk concept where the lender puts up the money and against collateral. Prudent lenders ask some reasonable ownership in the collateral, as they will not finance 100%. If they do so, lender will have no risk and nothing to loose if he/she misses payments. As a general rule, a loan-owner should never borrow 100%, and a good lender will never loan 100%. The type of equity depends on the asset, but a general rule is not less than 30-35% of its value.

Lenders want collateral for two reasons; the first is control. The second is that collateral provides the last source of repayment. Credit underwriting is a judgment call utilizing all of the above. Good lenders will not provide credit based on the strength of a single item, but may very well deny credit based on the weakness of a single item. As an example, our largest single reason for denial is lack of collateral equity, meaning people want to borrow too high a percentage of the value of the asset provided for security.

When or if you find a lender who is willing to advance funds when the risk is above acceptable levels – most often you have found a lender who charges excessive fees and/or interest, and is totally focused on the collateral. These lenders are normally referred to as "hard money lenders", they provide funds with little to no consideration of future, but they fully understand the concept of collateral. While most good lenders view collateral as the last source of repayment, these lenders view collateral liquidation as the first source of repayment. To accept a loan under these circumstances gambles collateral against chances of future success, while minimizing its probability due to excessive interest and related costs.

3.4 Loan Process

3.4.1 Introduction

The loan process is universally used within the agricultural lending world, but there is little uniformity from loan-to-loan or lender-to-lender. There are many ways a loan officer can analyze a loan. The process begins with establishing the character and ability of the borrower. The loan officer must meet with the borrower to assess the character of the borrower. The borrower is assessed on a subjective basis as the loan officer analyzes the operation, credit history, and general behavior of the borrower. If there is any indication of poor character, the loan officer assesses the borrower as a high risk.

After character is established, the farm is evaluated. The farm's management is evaluated subjectively as well. Virtually no two operations run the same way, so the analysis of farm management is relative to the operation and no number-ranking system can be uniformly employed in the analysis of all farming operations. Management would be deemed acceptable if there is an optimization of profit, employee satisfaction, employer satisfaction, and overall farm maintenance.

Then the loan officer must look at the current state of the market of the farmer's commodity, product, or service. The loan purpose is taken into account, as well as the repayment capacity of the borrower if the loan were to be issued. Collateral is then considered and evaluated based on the appraised value and the ability for the lender to gain possession of a first position lien on the collateral. Collateral is viewed as acceptable if its value is higher than the loan amount and if the collateral's useful life exceeds the term of the loan

The risks are taken into account as well as the lender's costs. Both help calculate the rate at which the lender will be covered from a possible default. Finally, basis points are added to that rate which gives the lender their profit. All of this is accumulated into the interest rate the lender charges the borrower. Once this is complete, the loan officer considers all the components of their research and analysis in order to make their decision. The decision process for a loan is a multi-attributed decision in which decisive variables are taken into account and analyzed through the eyes of the individual loan officer.

3.4.2 Loan Process/Analysis

In a study in 2005, three men collected data on 963 borrower-level lending relationships from commercial banks and FCS members in the Northwestern United States. The data included risk analysis from loan officers, data collection forms distributed to loan officers, and self-administered questionnaires completed by each loan officer. The information was used in a regression equation which showed the longer the relationship between borrower and lender, the lower servicing and monitoring costs were. The regression also provided that the more loan extensions made between the borrower and lender led to small decreases in costs. This shows the loan officer's relationship with the borrower is the foundation for the loan process.

After a relationship has been established and character has been analyzed, the loan officer must look at the applicant's credit risk. Credit risk is a rather subjective idea, but it is objectified as much as possible in the lending world. Many programs are available to analyze credit risk. Making sure an effective program is being used is essential. A study found that a member of the Farm Credit System was using a model that was intended to be used by commercial banks. Commercial banks operate in a different manner than the Farm Credit System. In comparing two different models, the study found the FCS bank was over capitalized, which meant they could be lending more money if they had been using the correct model, Credit Risk+. This model took into account agricultural aspects of credit analysis that the other model did not. Theoretically, if banks lend out more money, they make more money. So proper credit risk analysis has some major implications in the agricultural lending industry.

After credit risk is established, the loan officer analyzes the submitted financial information. The loan officer uses Microsoft Excel spreadsheets to calculate Debt Coverage Ratios, Current Ratios, Loan-to-Value Ratios, and Debt-to-Equity Ratios. These ratios are then analyzed in relation to cash flows within the applicant's business. The loan officer then puts this information into context with a hypothetical situation using a simulation model to determine the potential borrower's ability to repay the loan. Loans are priced by their interest rates, which are the costs of borrowing money. These interest rates are calculated through the loan officer's analysis. It has been shown there is an inverse relationship between the bank's size and their pricing. This means the larger the bank, the smaller the spread, and vice versa.

As per study which shows character, credit score, financial variables hat represents credit risk, lender experience, and the time spent working on the loan have significant influence on whether the loan is accepted as well as interest price charged on the loan. Higher credit scores, more favorable character analysis, decreased credit risk variables, higher lender experience, and less time spent working on the loan all led to a decreased interest rate.

One more thing considered in the loan process is collateral, or what will be taken by the lender if the borrower defaults on the loan. Real estate loans have good collateral because if the borrower defaults, the lender has the right to take possession of the real estate and sell it off to pay off the balance of the loan. Operating loans have a more difficult time with collateral because there is not always a tangible asset that can be collateralized. Even if there is a tangible asset, it is sometimes difficult to collect on it. For example, if crop inventory is used as collateral and there is a large drought, then the crop doesn't survive and the farmer has no profit to pay the loan back with. Therefore, the idea of risk contingent credit lending is a possible alternative to collateralized lending on operating loans. This risk contingent credit lending ties the interest rate to the risk of the commodity as well as the borrower, and has a built-in option or insurance which compensates the lender for a large, unexpected drop in the commodity's market price.

The loan process is a very long and involved process. Taking into account character, capacity, collateral, and capital, the loan officer can then determine conditions of the loan. This process melds art and science, the objective and subjective, in a manner to help loan officers formulate a decision whether to approve or reject a loan application. Using this process, I will make that decision.

3.5 Factors for keeping rules

3.5.1 Record Keeping

In order to provide what you need for decision making and what your lender will look for to approve your loan request, accurate and timely records are a must.

In today's age of computers there are few excuses left for not having complete, accurate, and timely records.

If the idea of record keeping is to take 12 bank statements to accountant at the end of the year and say "Do my taxes," then it needs improvement in internal record keeping which include:

- Accurate, detailed and individual listings of all assets, including: date purchased, make and model, cost, serial number or other description, expected useful life, and if pledged as security – who is the creditor.
- Accurate, up to date listing of all creditors, with balance outstanding, repayment terms, collateral, loan officer and complete contact information
- Balanced checking account, one that you know exactly how much money is in the bank, what checks have cleared, what checks are outstanding, etc. If you must access the bank via ATM and/or the Internet to find out your account balance, then your record keeping could use some serious improvement.
- Accurate and detailed record of all income and expenses. This must be broken down by the various categories in such a fashion that you can rapidly determine what you made from all the various sources, and what you have spent on all the various expense categories.
- Accurate and detailed records of all Accounts Receivable. When you sell something, until you have collected the money, it really doesn't do you a lot of good. You must know exactly who purchased what, when, and what the terms of repayment are. It is advisable to always get invoices signed. When you have Accounts Receivable, you are lending your money to the individual - it might be wise to evaluate the risk.
- Accurate and detailed listing of all Accounts Payable. This needs to include what the terms of repayment are.
- Production records. You need to know what you produced. This needs to include the number of acres, trees, cows etc as applicable for your operation.
- There are a number of good record keeping programs available for your PC. Classes are available at many local community colleges, and your tax accountant can make recommendations on how to keep accurate records. Use whatever means are necessary to ensure you have timely and accurate records.

3.5.2 Consumer Credit

The easy availability of consumer credit is a leading cause of problems with many agriculture producers. Just because a credit card company will say yes and give you a $25,000 line doesn't mean you should immediately take a trip to Vegas and run the card up. Don't utilize your credit cards for operating costs just because it's so easy. The interest costs are too excessive, and thereafter, all operating lenders will be very suspect of your prior practice. Do NOT use one card to pay another card; this is a game that will ruin your credit.

If you are taking more than 90 days to pay off your credit cards (from the date of the advance to the date that advance is totally paid off), then you may want to examine your credit spending habits. If you are past due with consumer credit, the first step in finding your way out of this trap is to throw away the shovel you are using to dig your financial grave with. The next step after you stop charging is to pay more than the minimum balance. There are a number of consumer credit counseling services available, but avoid those who are going to compromise your debt. This will almost always end up with larger problems than started with details related to consumer credit histories as viewed by lenders which covers:

- ❖ Excellent Credit – high score, NO derogatory, very limited occasional 30 day delinquency.
- ❖ Good Credit – NO derogatory, minor but non reoccurring occasional delinquency.
- ❖ Average Credit – Any derogatory is resolved, some occasional delinquency but no reoccurring 60 or over delinquency, all of which is explainable.
- ❖ Poor Credit – minor unresolved derogatory and/or multiple repeating delinquencies including accounts 60 and 90 over.
- ❖ Adverse Credit – multiple and/or significant unresolved derogatory accounts (judgments, collections, tax liens, bankruptcy or other public records), and/or multiple continuing delinquencies.
- ❖ Excellent, Good, and Average credit history can expect to receive financing, assuming there are no other weaknesses or problems. Regardless of the collateral, if you have poor and/or adverse consumer credit history, you should not expect to receive credit under normal circumstances and terms.

3.6 Determining Cash Flow

This is a primary planning tool, and will aid you in purchase or lease decisions. Generating sufficient cash from sustainable operations to make your payments is the only acceptable long-term method of operating. It follows that the annual cash flow is determined by using following items:

- ❖ Gross wages,
- ❖ Interest Income,
- ❖ Other non-business revenue,
- ❖ NET Schedule C earnings (or subtract net loss),
- ❖ NET Schedule E earnings (or subtract net loss),
- ❖ NET Schedule F (Farm) earnings (or subtract net loss),

- ❖ All of the above is normally totaled on your tax return as "total income" on row 22 of your tax return (watch out for capital gains – these should not normally be included as they are not reoccurring).
- ❖ Depreciation expense from Schedules C, E, & F,
- ❖ Interest expense from Schedules C, E, & F,
- ❖ Capital Lease payments.
- ❖ From the above subtotal, SUBTRACT all of the following:
- ❖ Income taxes, including self-employment, and state taxes if applicable,
- ❖ Family living expenses (without any debt payment)

Now what left is gross cash available to service debt as it is the amount needed which should be greater than total of all debt payments (including capital lease payments) for the year. If you have lines of credit, only consider their interest expense in the debt repayment total. The greater the margin, the better; most lenders want to see 120-135% "coverage", meaning for each dollar of debt payment due annually, you will have from $1.20 to $1.35 in cash available.

The smaller this margin, the greater the risk, so remember that history is the window of the future and when the operation is static in size and margin is less than 115%, then you must be very careful about any additional debt. If margin is consistently below 100%, you are headed for a disaster once your lender catches up to the information curve.

R's of Credit-Principles of farm credit

Credit is based upon belief, confidence, trust and faith. The loan is based upon the confidence of borrower's future solvency and repayment. Credit means ability to command the other's capital in return for a promise to re-pay at some specified time in future. Besides, credit is the combination of "ability to borrow" and "willingness to borrow", infact credit is an individual's borrowing capacity, often being considered as an "economic good" to be produced, managed and marketed.

There are three basic considerations, which must be taken into account before a lending agency decides to agency decides to advance a loan and the borrower decides to borrow:

- ❖ returns from proposed investment
- ❖ repaying capacity
- ❖ risk bearing ability of borrower

Returns from investment: First R of credit

The returns from an investment, serves as the first test of credit, which has great significance to both creditor and borrower. It requires that both borrower and lender are satisfied about the returns from credit which cover the principal and interest. Furthermore, the basic question pertaining to returns analysis is whether or not the use of credit generates adequate income and is most profitable use. Thus, even though the use of credit may be profitable it should also be examined whether or not it is the most profitable. Similarly, the examination of returns from an investment in terms of generating adequate incomes to compensate for the contribution of family labour and management as well as building owner's equity is also essential. Hence, overall profitability of a farm business must be evaluated to assess the possibility of earning income from most profitable enterprise to compensate the loss from another.

The problem of determining the most profitable use of capital is a part of decision making which involves

- the selection of most profitable enterprises (product-product relationship),
- determining the most economical production techniques (factor-factor relationship)
- determining the size of each enterprise (factor-product relationship).

However, the following points may be kept in mind while calculating the expected returns from the borrowed funds.

- Estimate the gross returns by multiplying average yields with its corresponding expected average prices, the conservative prices should be used for safety purpose. These gross returns should be worked out both with and without borrowed funds.
- Estimate the total cost both with and without borrowed funds, these costs should be slightly on higher side to take into account the risk.
- Estimate the additional cost and additional returns from the investment, do not use average cost and average returns.

The use of credit becomes an economically sound proposition, if the net cash income is more due to the use of borrowed funds, with a sufficient margin for income variability.

Repayment capacity: Second R of credit

It should also be taken into consideration while extending/borrowing a loan. In fact, it is not only sufficient for a loan to be productive rather it should also generate adequate returns so that loan installments can be repaid. It is quite possible that a loan may be productive but may not generate adequate income after meeting the family and farm expenses to regularly pay the loan installments.

The repaying capacity is the amount of money that a farm family would be able to spare from their total earnings so as to repay the loan after meeting his farm and family expenses. Ability to repay a loan is influenced by the income generating capacity of the farm business, off farm earnings, the liquidity of the farm as reflected by the balance sheet and the cash flows on the farm (with due consideration for farm and family obligations). Furthermore, the ability to repay may be influenced by numerous factors but willingness to repay a loan is quite essential. In short run, the current assets must be able to repay the current liabilities.

However, in long run the sufficient income must also be available (after meeting operating expenses, family living expenses and expenses for the farm-firm growth) so that debt obligations can be repaid. In this contest, the cash flow analysis of a farm business depicts the complete picture of repayment ability, i.e., the period of cash inflows so that loan terms and payment dates can be tailored with it. For maximization of the net returns from an investment, the credit is to be used upto the point where additional (marginal) net income equals to that of additional (marginal) cost. Therefore, the repayment of principal and interest depends upon the type of loan taken by the farmers, viz., self-liquidating and non-liquidating or partially liquidating loans.

Self-liquidating loans

These are loans to acquire goods or services that are completely used up in one production season or in annual production process. These are the short-term loan (operating expenses) which become a part of working expenses in the single production process. In estimating repaying capacity the borrowed funds are not deducted as costs are included in working expenses. The repaying capacity for such loans should be determined as below:

Repaying capacity = (Gross income including off-farm income) - (living expenses + working expenses excluding proposed loan + taxes and L.I.C. premiums + other loans and repayments due).

Non-liquidating loans

These are loans where resources acquired are not expended or consumed up in a single production process, i.e. the acquired resources are consumed over a number of years. Such loans do not completely become a part of the first year's costs and returns from such investments are spread over a number of years. A loan for the purchase of tractor or land reclamation is an example of non-liquidating loan, i.e., all the medium and long term loans are non-liquidating loans. These loans may contribute indirectly to the repayment capacity by enabling the farmer to produce more net income than otherwise would be possible without the use of such resources. The repaying capacity for such loans is worked out as:

Repaying capacity = (Gross income including off farm income) minus (working expenses including seasonal loans + living expenses +taxes and LIC premiums + repayment of other loans due).

Poor debt-servicing capacity can be due to:

- Small operational land with inadequate infrastructures on these farms ultimately results into the lower gross returns.
- Lack of appropriate avenues for off-farm income in the area/region.
- Poor yield of crops due to factors such as poor land, inadequate/erratic rainfall, salinity/alkalinity problems, disease and pest infestations, low adoption of improved technology, etc. Similarly, rearing of low yielder milch animals, small herd size, lack of adequate feed and fodder, veterinary facilities, in-efficient milk marketing etc. may result lower returns from milch animals.
- Sale of crop produce during the post harvest months (also through in-efficient channel), distress sales and poor market infrastructures.
- High production costs due to high input prices (also due to in-efficient purchase of farm inputs).
- Diversion of loans for unproductive purposes, and High cost of living.

Risk bearing ability-Third R of credit

Risk bearing ability, the third R of credit, determines the quantum of credit which can be safely used by the farm-firm. It means the ability of borrower to withstand the unexpected low incomes, unpredictable losses and expenses and to continue the farming. It provides the "last line of defence" in the use of credit. It is quite possible that a loan may be productive and may also generate adequate repaying capacity but borrower may not be able to afford the shocks of probable financial losses due to poor/inadequate risk-bearing ability. The assessment of risk bearing ability is, therefore, essential since both returns and repayment capacity are based on average estimates of production, prices and costs, which seldom hold true. Farming is exposed to the many natural hazards such as, attack of insects, pests and diseases and also to the price fluctuations. Instability in farm income is more common as compared to the stable farm income. Consequently, most of the farmers are risk- averter rather than the risk preferer. Hence, the variability in gross incomes has to be accounted for before arriving at fairly stable and also reliable estimates of the level of repayment capacity. The stable repayment capacity, based on deflated gross returns, may be worked out to account for risk-bearing ability of the borrower. For this purpose, the general variability coefficients of the area may be considered due to non-availability of these coefficients for an individual borrower in developing countries. From deflated gross income, deductions should be made for farm and family expenses and the loans due. The balance would indicate the repayment capacity for the relevant period.

Stable repayment capacity = (Deflated gross returns minus working expenses plus family living expenses plus old debt payments plus L.I.C. premium, etc.)

The stable income or financial position though determines the borrower's ability to bear the financial shocks yet the risk bearing ability of a borrower depends upon the following factors:

- Ability and willingness to save.
- Ability to borrow, i.e., credit worthiness of the borrower as a person, especially in bad times.
- Ability and willingness to adjust and withstand the adverse conditions, i.e., reducing both operating and living expenses in bad periods.
- Equity and net worth, the backbone of risk bearing ability. The risk bearing ability can be enhanced by certain measures such as:
  - Taking crop, livestock and other insurances.
  - Adoption of financial strategies (e.g. internal cash or asset rationing, internal and external credit rationing and reducing farm and family expenses.)
  - Adoption of suitable marketing strategies (such as hedging, forward contracts for sale of farm products and purchase of input supplies to reduce price risk).
  - Adoption of suitable production strategies (such as flexible production programmes, use of plant protection, weedicides and other farm practices, growing less risky or more stable farm enterprises, diversification of farm production programmes).
  - Building up of owner's equity or net worth through savings and personal credit through fair dealings.

4

Sources of Agricultural Finance

4.1 Introduction

Finance has been recognized as the life blood of all economic activities. Like all other producers, agriculturist also needs credit. According to an old proverb "Credit supports the farmers as the hangman's rope supports the hanged." This statement is fully true in the context of Indian farmers. Thus, for stimulating the tempo of agricultural production, an adequate and timely credit, is most essential.

Generally, in underdeveloped countries farmers cannot expect their credit needs to come from savings. It is so because their income from farm operations is sufficient to provide minimum necessities of life. Therefore, they have to rely upon outside finance. In olden days, rural debt was considered as an unmixed evil, but now time has changed altogether. Modern agriculture is a costly affair. Credit is needed to adopt new farm technology resulting in ushering of green revolution. In India, it has two fold necessity. Firstly crop productivity is very low due to traditional methods of cultivation and secondly, there is an urgent need to enhance agricultural production to get self sufficiency and to save valuable foreign exchange. In short, effective arrangements are needed to provide credit facilities so that agriculturist may adopt better techniques of production.

The different studies conducted show a strong positive relationship between agricultural growth and availability of credit. Broadly, credit in agricultural sector may be divided into short-term loans to meet the input expenses and medium and long-term loans to facilitate the development of fixed farm assets such as land. This gap arises in relation to static or dynamic production function. Under a static functioning, the level of input use per hectare of cropped area being constant, the year to year variation in the amount of credit reflects the changes in input prices.

Under a situation of diminishing returns, however, increasing input use is required to maintain the same level of output. The supply of credit related to static production conditions will not contribute to increase in output, although the withdrawal of it might lead to a decline in conditions will not contribute to increase in output, although the withdrawal of it might lead to a decline in output. Under dynamic functions, credit requirements would rise from year to year even if input prices remain constant. The growth in credit under such dynamic conditions would lead to increased output. In the same way the investment credit too would lead to an improvement in the production potential of the farms through the process of net capital formation.

The agrarian history from Rome to Scotland is that agricultural credit is an essential. Neither the condition of the country nor the nature of land tenures nor the position of agriculture affect one great fact that farmers must borrow. Agricultural credit, thus, in a practical sense, is a nucleus of the system of farm operation. It provides flow to the system averting ruins which would have occurred due to the lack of monetary capacity of farmers. Thus adequate and timely credit to the farmer is, vital and indispensable for the rehabilitation and progress of agriculturists. In underdeveloped countries, agriculture assumes even more importance. Farmer's inability or least limited ability to save does not allow him to finance his pursuits and raise better production from his farms. Agricultural credit through institutional channels is the only way to break agricultural stagnation. Private funding agencies play a limited role keeping in view the larger public interest.

4.1.1 Features

In our country, agricultural finance has the special features which are discussed below in detail:

- Risks in Agriculture: In agriculture sector, it is difficult to foresee risks and uncertainties. A farmer has to face numberless risks and uncertainties as droughts, floods etc. It may cause considerable damage to the farmer. Moreover, agricultural produce tends to deteriorate in storage due to lack of proper storage facilities to hold back surplus when supply exceeds demand. It leads to further difficulties. Thus, with so much uncertainties, agriculture has always been a risky affair to be handled by the commercial banks and insurance companies.
- Difficulties of Co-operation in Agriculture: In agricultural sector, there is a very little scope of cooperation. It is so because, farmers are mostly individualistic and are suspicious of co-operating with each other for a common purpose. This creates difficulties to the farmers in getting cheap credit.
- Economic Lags in Agriculture: In agricultural production process, there is a long interval between the reward and effort specially during the period

when costs are incurred. During this period, demand for agricultural produce may change upsetting the financial adjustments of the farmers. In this way, farmers have to bear another uncertainty. This becomes an excuse for credit supplying agencies to refuse credit for farm operations.

- Credit for Consumption Purpose: Indian farmers require credit not only for production purposes but also for consumption purposes. In the case of crop failure, small farmers need credit which they spend on consumption requirements. Moreover, Indian farmers are accustomed to spend beyond their means on social and religious functions. In addition to all this, litigation is another important non-productive requirement for funds.
- Small Size of Farm: In India, size of farms is very small in comparison to the amount of labour employed and the extent of the capital invested. Moreover, there is no control over the yield and the quality of the produce. Thus, there is a lack of security to be offered for loans.
- Lack of Proper Securities: The large farmers have their own resources which enable them to raise funds from the credit institutions. Small farmers find it extremely difficult to raise credit for their needs. It is due to the reason that small farmers neither possess proper securities to pledge against loans, nor they have adequate repaying capacity. As a result, small farmers are forced to go to the money lenders.
- Complex of Many Industries: Agriculture is an industrial complex of varying types of production and marketing. The size of holdings and forms of land tenure differ from one area to another. These differences create different types of complex relations between the farmers which makes financing of agricultural sector relatively difficult.

4.1.2 Criteria for Agricultural Credit

In any scheme of agricultural finance, there must be some criterion on the basis of which the superstructure of various agencies has to be built up. Since the farmer's needs for finance and the methods of operation differ widely from non-agricultural financing, it is more necessary to keep certain criteria in mind. According to Mr. Louis Tardy, the various criteria of a suitable type of credit may be set down as follows:

- To be granted for a sufficiently long period, commensurate with the length of the operation;
- To be granted at a low rate of interest;
- To be adequately secured, in order, more particularly, to avoid any abuse of credit facilities, but the security should not necessarily be material; it should be in the form of a personal credit secured mainly by the borrower's moral standing and farming ability;

- be adapted to the average yield and capacity for repayment of the farms, particularly during periods of economic depression;
- To be placed in the hands of institutions, the directors of which have received special training; and have actual banking experience.

The following criteria may also be added.

- There should be possibility of extension of the term of credit in the event of crop failure;
- It should be made locally available without delay; and
- The agency engaged for the purpose should be able to find out facts about the borrowers readily.

4.2 Institutional source of agricultural finance

The general policy on agricultural credit has been one of progressive institutionalization aimed at providing timely and adequate credit to farmers for increasing agricultural production and productivity. Providing better access to institutional credit for the small and marginal farmers and other weaker sections to enable them to adopt modern technology and improved agricultural practices has been a major thrust of the policy. National Bank for Agriculture and Rural Development is an apex institution established in 1982 for rural credit in India. It doesn't directly finance farmers and other rural people. It grants assistance to them through the institutions described as follows:

4.2.1 Rural Co-Operative Credit Institutions:

The rural credit system in the country has undergone radical changes in respect of focus, structure and approach over the years. Prior to the institutionalization of credit, the farmers were mainly dependent on the non-institutional credit especially on private money-lenders, who failed to provide the farmers the necessary and timely credit at appropriate cost. In order to overcome these hurdles and to supply the farmer's adequate and timely credit the institutionalization of credit was started with the establishment of cooperative societies with the enactment of cooperative societies Act in 1904. Till 1969 cooperatives were virtually the only institutions for dispensing rural credit until the nationalization of 14 commercial banks, to improve the flow of credit to rural households.

Both the cooperatives and commercial banks have made substantial progress over the years in providing credit to agriculture under "priority sector" advances as per the guidelines of Reserve Bank of India. The cooperative credit structure has two wings namely production credit (short-term credit structure) which comprises of Primary Agricultural Credit Societies (PACS) at the base level, District Central Cooperative Banks (DCCB) at the intermediate level and State Cooperative Banks (SCB) at the apex level. The investment credit (long term credit structure) comprises of State Cooperative Agriculture and Rural Development Banks (SCARDBs) at the apex level and Primary Cooperative Agriculture and Rural Development Banks (PCARDBs) and their branches at the base level.

Indian Rural Cooperative System is one of the largest rural financial systems in the world with around one lakh societies serving the credit and agricultural input needs of around 130 million farmers. Though cooperatives are 106 years old in India, no rural cooperative credit institution features among the top three hundred in the world. The rural cooperative credit system in India has been living with a lingering affliction of body and mind which has left it wasting away. The cooperative system has been beset with a multitude of administrative, structural, functional, economic and socio-economic problems. Dormant membership, lack of active participation of the members in the management, lack of professionalism, absence of corporate governance, politicization and bureaucratization and ageing and unenthusiastic human resources have weakened the cooperative edifice. The financial cooperatives in India have mainly suffered from 'seven deadly sins'

- Reliance on external funding rather than the deposit mobilization
- Lack of meaningful financial information in balance sheets
- Inappropriate interest rates on deposits and credit
- Poor public image due to absence of physical facilities, human resources and marketing
- Undisciplined financial management in five key areas- delinquency control, adequate loan loss provisioning, sufficiency of institutional capital, maintenance of adequate liquidity reserves and proper asset-liability management.

Cookie-Cutter analysis – credit decision based on multiples of the person's shares or savings and not on the capacity, past credit history, and willingness to repay. Social philosophy over common business sense.

The issues pertaining to cooperative system have been examined by a plethora of committees, task forces and working groups starting from Macglean Committee in 1914 to the recent Task Force on Revival of Rural Cooperative Credit Institutions in 2004. For a movement of vital importance to the well-being of the nation to suffer without a lasting remedy over such a long period of time is indicative of 'some elusive truth remaining hidden from a succession of well-meaning seekers and policy-makers'.

Rural Credit cooperatives are the oldest and most extensive form of rural institutional financing in India. The major thrust of these cooperatives in the area of agricultural credit is the prevention of exploitation of the peasants by moneylenders. The rural credit cooperatives may be further divided into short-term credit cooperatives and long-term credit cooperatives.

Short Term Credit Cooperatives

The short-term credit cooperatives provide short-term rural credit and are based on a three-tier structure as follows:

(a) Primary Agricultural Credit Societies (PACs):

Primary Agricultural Credit Societies (PACS) occupy a predominant position in the co-operative credit structure and form its base. A PACS is organized at the grass roots level of a village or a group of small villages. It is this basic unit which deals directly with the rural (agricultural) borrowers, gives those loans and collects repayments of loans given. It serves as the final link between the ultimate borrowers on the one hand and the higher financing agencies, namely the SCBS, and the RBI/ NABARD on the other hand.

As such, the health and strength of the co-operative credit movement depends crucially upon the health and strength of these societies. But, despite much official effort and support, and numerical expansion of the PACS in membership, working capital, loans given and other activities, their health and working leave much to be desired.

At the end of June 1989 there were 87,000 PACS. 'These societies covered about 90% of 5.8 lakh villages. Their membership of 9 crores covered about 65% of the total estimated population of about 14 crores of rural households. More than half of the members of PACS are persons of small means—small fanners, agricultural laborers and rural artisans and about 25% of them belong to scheduled castes and tribes.

All these are very strong features of the co-operative credit organization, as it alone provides the vast network of institutional credit agencies throughout the length and breadth of the country of sub continental size, covering the large bulk of Indian villages. And yet the organisation is not fulfilling its role adequately due to several weaknesses.

The working capital of the PACS is derived mainly from borrowings from the CCBs and in smaller proportions from owned funds and deposits. That the PACS have failed to attract deposits is not so much a reflection of low savings habits of the rural population as a reflection of the availability of better assets to rural savers in terms of both rate of return and riskiness.

The richer farmers either invest their saving directly in physical assets or lend them directly as agriculturist moneylenders to others in the village or hold them in deposits with banks or post offices. Only the residual savings of those households come to the PACS who want to borrow later from these societies or who want to wield influence in the actual working of the societies. To attract more deposits, these societies must be in a position to pay higher interest rates on their deposits than offered by other institutions and must inspire greater confidence regarding safety of deposits with them. For this, the societies must be managed efficiently as well as honestly.

PACS are organized at the village level. These societies generally advance loans only for productive purposes. The main objective of a PACS is to raise capital for the purpose of giving loans and supporting the essential activities of the members such as supply of agricultural inputs at cheap price, improving irrigation on land owned by members, encourage various income-augmenting activities such as horticulture, animal husbandry, poultry etc. In India, around 99.5 percent of villages are covered by PACs.

(b) District Central Cooperative Banks:

With the introduction of cooperative credit societies Act in 1904, no provision for the organization of DCCBs was made in the Act which provided for the organization of primary societies only. The original scheme of cooperation in India did not contemplate the organization of federal societies. Hence most of the PACS in the district were starved of money. Further they could not attract deposits whereas the demand for loans from the members was increasing. Subsequently the DCCBs came in to existence to coordinate and supplement resources to PACS. In 1912 the crop credit societies 1904 Act was replaced with cooperative societies Act with an objective of organizing DCCBs. Consequently, a number of DCCBs have been registered in several areas to mobilize funds mainly in the form of deposits and also to borrow money from higher financial institutions in order to meet the financial needs of affiliated primary societies.

The DCCBs occupy an important place in the three-tier cooperative credit structure. The DCCB is a federation of PACS in the district. They form an important link between the Apex cooperative Bank and the PACS. The DCCBs are at the district level, directly connected with PACS at the base level. As a district level institution, the DCCB is responsible for the promotional and developmental activities in the state. A DCCB in addition to meeting the credit requirements of affiliated PACS also lends a helping hand to other types of primary cooperatives like marketing, consumer, weavers cooperatives, cooperative sugar federation, cooperative spinning mills etc., in the district. In mobilizing resources, in overseeing the lending procedures, in supervising the recoveries and above all in educating the members, the DCCBs are expected to provide leadership to PACS in the district.

The working of DCCBs in India has taken different strides due to many factors such as booms and depressions, world wars and policy developments. Several committees were appointed from time to time to strengthen the rural credit structure in the country. The important committees appointed by Government of India, RBI and others were The Maclagan Committee (1914), Cooperative Planning Committee (Saraya Committee) 1945, All India Rural Credit Survey Committee

(AIRCSC) 1954, All India Rural Credit Review Committee (B. Venkatappaiah Committee) 1969 and Task force on rural credit (2004) these committees suggested useful recommendations to strengthen DCCBs in the country.

The Maclagan Committee (1914)

Sir E.D. Maclagan committee was appointed in the year 1914 to review the progress of cooperative movement and to suggest measures to strengthen the cooperative movement. The important recommendations of the committee were, confining the business of cooperatives to members only, entrusting the work of supervision of cooperatives to Cooperative Unions, providing freehand to management in assessing and sanctioning the credit limits, restricting the area of operations of DCCBs to a district or tehsil, linking of borrowings to shareholdings of societies, gradual discouragement of individual members, provision of production oriented loans and measures for adequate deposit mobilization.

Cooperative Planning Committee (Saraya Committee) 1945

The CPC enquired in to the possible ways of improving the financial strength of DCCBs and suggested for the coverage more population under the cooperative fold. The other recommendations include entrusting supervision responsibility to DCCBs wherever the cooperative unions are not available.

All India Rural Credit Survey Committee (AIRCSC) 1954

The main recommendations of AIRCSC were, linking of borrowings with members share holdings, share capital contribution by Apex Bank, maintenance of Agricultural credit stabilization fund and contribution of financial assistance by the RBI to the DCCBs in the form of ST & MT credit limits etc.

All India Rural Credit Review Committee (B. Venkatappaiah Committee) 1969

The important recommendations of the committee include provision of increased share capital assistance to weak DCCBs, provision of training facilities to the staff, re-organization of weak DCCBs at the rate one for each district with Rs.1 crore business, entrusting the supervision responsibility to DCCBs and maintenance of financial discipline.

In order to extend banking facilities to the rural areas, the DCCBs open branches within the jurisdiction by simply passing a resolution in the Board, which is a privilege given by the RBI. At present the DCCBs are competing with commercial banks in mobilizing deposits and also in marketing of banking services in rural areas as well as in urban areas. This branch arrangement helps them to improve their contacts with the PACS. Further this also helps to tap the savings of rural people through deposit mobilization.

These cooperatives are organized at the district level. The PACS are affiliated to the District Central Co-operative Banks (DCCBs). DCCBs coordinate the activities of district central financing agencies, organize credit for PACs and carry out banking business.

Administrative Set Up of DCC Bank

The central banks have pyramid type of administrative set up and are managed by Board of Directors consisting of 12 to 15 members. In order to streamline the work of the bank several sub-committees like Executive committee, Law & Liquidation committee, Industrial committee, Loan committee, Building committee are formed and the jurisdiction of these committees are determined. All the decisions pertaining to the disbursement of credit, deposit mobilization, investment of capital etc, are taken by the chief executive in consultation with the approval of the board or concerned sub-committee. The president or chairperson is the head of the institution and the vice-president or vicechair person performs the duties of the president or chairperson in his absence.

The Chief Executive of the bank is the General Manager/Manager/ Secretary. He undertakes all the functions of the bank with the help of Assistant Managers/ Assistant Secretaries and the subordinate staff. The bank has a separate set up for branches. Besides the divisional managers in the taluk places and branch in general, there are inspectors and other subordinate staff for the day to day operations of the branch or division.

The key decisions in connection with the man power planning such as recruitment, promotion, transfer are taken by the BOD on the basis of the information supplied by the branches and other sections in the head office. As for financial planning in the area of deposit mobilization, advance of loans, recovery of loans, investment etc, are concerned, all most all policies laid down by the NABARD and State Government are being followed.

The bank collects information from its branches and divisions in the form of Daily, weekly, Monthly and quarterly reports for the future planning and development. The bank also provides reports to RBI, NABARD on the basis of these reports.

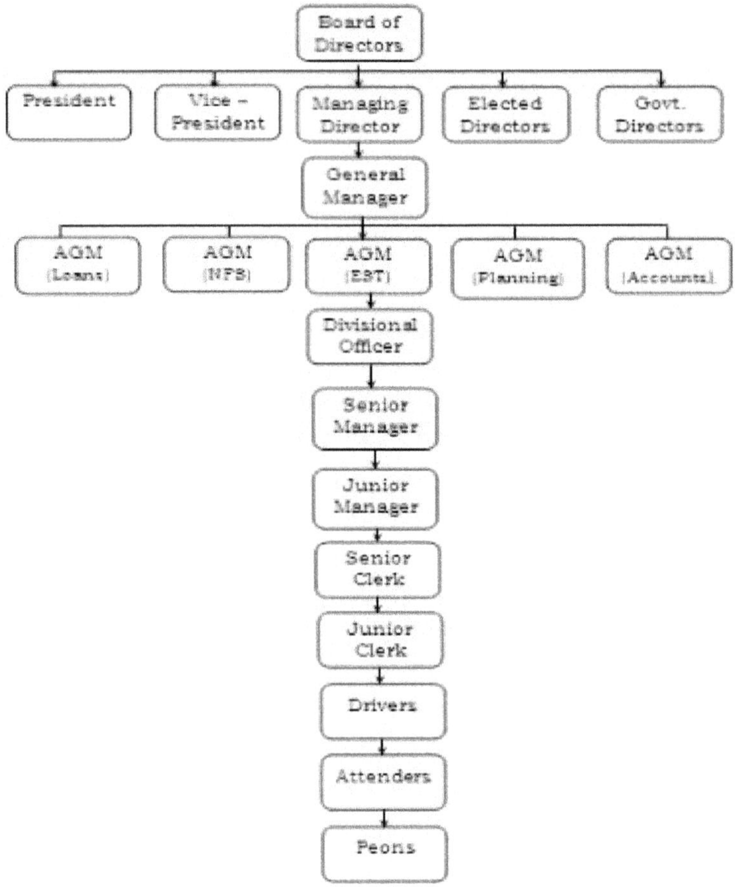

Fig Organization Chart

Sources of Finance

The resources of DCCBs are classified in to two categories; namely

- ❖ Owned Funds
- ❖ borrowed funds

The main components of owned funds are share capital, Reserve Fund & Entrance Fees. The borrowed funds consist of deposits, borrowings from RBI, NABARD, SCBs & other DCCBs.

Share Capital - The share capital forms an integral part of owned funds. An individual or a Primary level society has an obligation to buy at least one share of the bank for enjoying the rights of membership. The PACS/PLS, the State Government and other special Agencies like SFDA are 'A' class members, while the individuals are 'B' class members.

State Government Participation - The RBI has set up a fund called National Agriculture Credit (Long Term Operations) for the purpose of making advances to state governments for enabling them to contribute the share capital to DCCBs. The NABARD has been entrusted with the above function.

Deposits - Deposits are an important tool for resource mobilization. Higher deposits will enable the banks to reduce the dependence on external capital. At present it is more important because of the fact that the NABARD has linked the concessional finance to banks to deposit mobilization performance of the DCCBs.

Borrowings - It is a normal feature of the bank to borrow funds as and when internal funds are not sufficient. The volume of borrowings depends upon the deployment of funds and repayment performance. Further the volumes of borrowings are limited to 10 times the paid up share capital + RF or as permitted by the RCS from time to time. The banks are borrowing regularly from NABARD.

Cash Credit Limits - The DCCBs are permitted to draw the advances from the SCBs, as and when the demand for loans arises from the members. This is termed as cash credit facility. The RBI also provides rediscounting facility or making advances against 'cooperative papers'.

(c) State Co-Operative Banks

The state co-operative bank is a federation of central co-operative bank and acts as a watchdog of the co-operative banking structure in the state. Its funds are obtained from share capital, deposits, loans and overdrafts from the Reserve Bank of India. The state co-operative banks lend money to central co-operative banks and primary societies and not directly to the farmers.

The DCCBs are affiliated to State Co-operative Banks (SCBs), which coordinate the activities of DCCBs, organize provision of finance for credit worthy farmers, carry out banking business and act as leader of the Co-operatives in the States.

Functions

- They function with the rule of "one member, one vote" and function on "no profit, no loss" basis
- It performs all the main banking functions of deposit mobilization, the supply of credit and provision of remittance facilities
- It provides financial assistance to the people with small means to protect them from the debt trap of the moneylenders
- It is engaged in tasks of production, processing, marketing, distribution, servicing and banking in India
- It supervises and guides affiliated societies
- Mobilization of funds from their members

- Advance loans to the members
- Rural financing for farming, cattle, milk, hatchery, personal finance, etc.
- Urban financing for Self – employment, Industries Small scale units, Home finance, Consumer finance, Personal finance
- List of Advantages of Cooperative Banks in India:
- Easy to form
- No obstruction for membership
- Limited liability
- Service motive
- Democratic management
- Stability and Continuity
- Economic operations
- State patronage

Weakness

- They are too small to be economical and viable; besides too many of them are dormant, existing only on paper
- Co-operative banks are not doing well in all the states; only a few accounts for a major part of their business
- These banks still rely very heavily on refinancing facilities from the government, the RBI, and NABARD
- They suffer from dangerously low or weak quality of loan assets, and from the highly unsatisfactory recovery of loans.
- They do not look like banks and do not inspire confidence in the potential members, depositors, and borrowers.
- Most of the Co-operative banks are suffering from a lack of professional management.
- Except for some Co-operative banks, technological development in Information Technology or computerized data management is conspicuously absent.
- As there is no formal system of corporate governance in co-operative banks, many banks have become the hotbed of political patronage, unscrupulous financial practice, and gross mismanagement.
- Another problem arises out of the duality of control over them i.e. these banks are organized under the dual control of RBI and as well as respective state government.
- They unduly depend on government capital rather than member capital.

Problems of State Co-operative Banks

However, concerns regarding the professionalism of urban co-operative banks gave rise to the view that they should be better regulated. Large co-operative banks with paid-up share capital and reserves of Rs.1 lakh were brought under the purview of the Banking Regulation Act 1949 with effect from 1st March, 1966 and within the ambit of the Reserve Bank's supervision. This marked the beginning of an era of duality of control over these banks. Banking related functions (viz. licensing, area of operations, interest rates etc.) were to be governed by RBI and registration, management, audit and liquidation, etc. governed by State Governments as per the provisions of respective State Acts. In 1968, UCB's were extended the benefits of deposit insurance.

Towards the late 1960s there was debate regarding the promotion of the small scale industries. UCB's came to be seen as important players in this context. The working group on industrial financing through Co-operative Banks, (1968 known as Damry Group) attempted to broaden the scope of activities of urban co-operative banks by recommending these banks should finance the small and cottage industries. This was reiterated by the Banking Commission in 1969.

The Madhavdas Committee (1979) evaluated the role played by urban co-operative banks in greater details and drew a roadmap for their future role recommending support from RBI and Government in the establishment of such banks in backward areas and prescribing viability standards.

The Hate Working Group (1981) desired better utilization of bank's surplus funds and that the percentage of the Cash Reserve Ratio (CRR) & the Statutory Liquidity Ratio (SLR) of these banks should be brought at par with commercial banks, in a phased manner. While the Marathe Committee (1992) redefined the viability norms and ushered in the era of liberalization, the Madhava Rao Committee (1999) focused on consolidation, control of sickness, better professional standards in urban co-operative banks and sought to align the urban banking movement with commercial banks.

A feature of the urban banking movement has been its heterogeneous character and its uneven geographical spread with most banks concentrated in the states of Gujarat, Karnataka, Maharashtra, and Tamil Nadu. While most banks are unit banks without any branch network, some of the large banks have established their presence in many states when at their behest multi-state banking was allowed in 1985. Some of these banks are also Authorized Dealers in Foreign Exchange.

Long-term credit Cooperatives

These cooperatives meet long-term credit of the farmers and are organized at two levels:

(i) Primary Co-Operative Agriculture and Rural Development Banks

These banks operate at the village level as an independent unit. Despite declining share of agriculture in the nation's Gross Domestic Product (GDP), agricultural sector has been playing an important role in Indian economy. Interestingly, the number of people engaged in agriculture has not come down drastically. Majority of population still depend on agriculture for their livelihood. Hence, rapid growth of the agricultural sector has become necessary to achieve faster and inclusive growth. There are a number of reasons as to why the agriculture sector has not been growing at a faster rate. One of the problems faced by the agricultural sector has been lack of availability of timely credit. There are a number of institutions providing credit to country agriculture such as commercial banks, Regional Rural Banks and a host of cooperative institutions. The cooperative institutions provide short term as well as long term loans for various purposes.

The co-operative banking sector plays an important role in providing credit to agricultural sector. The structure of the co-operative banking sector in India is twofold. The urban areas are served by urban co-operative banks and the rural areas are served by two distinct sets of the institutions extending short-term and long-term credit. The short-term co-operative credit institutions have a three-tier structure comprising State co-operative banks at the apex level, district central co-operative banks at intermediate level and primary agricultural societies at the base level. The long-term co-operative credit institutions have, generally, a two tier structure comprising the State co-operative agriculture and rural development banks (SCARDBs) at the state level and primary co-operative agricultural and rural development banks (PCARDBs) at district or block level. PCARDBs are required to meet all the long term credit needs of the farmers. Generally, these societies are formed to free the weaker sections of society from economic exploitation. For example, poor farmers start cooperative credit societies to get loans at lower rate of interest and to free themselves from money lenders.

Small producers in order to save themselves from the middlemen and to get reasonable price for their products start marketing cooperative societies. People from low income strata start consumer cooperative societies to get the essential commodities at reasonable prices. The main objective of cooperative societies is "Society is for individual and individual is for society'. Here importance is given to the service rather than profit. Under this system, people with limited means are benefited much. Infect, the back has been successful in preventing farmers from taking help of money lenders/ an account of availability of long term credit by PCARDSBs, the farmers have been able to invest huge sum of money in agriculture. Some of the advantages always follow the bank such advantages are rate of interest are low on secured loan amount, loans given to particular business only, loan process consumes lot of time, unlimited negotiation are required, proper updates, some scheme are most worth. Co-operatives account for a relatively small share in the bank-dominated Indian financial system; however, given their geographic and demographic outreach, they hold a key position in the system. Geographically, co-operatives have been instrumental in extending formal financial services to villages and small towns in India. Demographically, these institutions have enabled access to financial services to low and middle-income groups in both rural and urban areas.

Finally cooperative for every village and every village should be covered by a cooperative". Cooperation as an organization refers to an association formed for the purpose of mutual benefits. As was said by The Royal Commission on Agriculture in India, The Indian cooperative movement was initiated by the Government. It spread and diversified with the encouragement and support of the Government. Its present condition is also to a great extent because of the intrusive involvement of and interference by the Government for development of agricultural sector.

Over the years, the primary cooperative agriculture rural development banks have lent amount to agriculture. These banks have contributed modern techniques, machineries of agriculture in the state. These banks by lending long term loans have externally competitions. Its coverage among farmers to undertake investment in agriculture. The banks have been facing a number of problems in the purchase of landing, verifications of documents of the borrowers and ensuing paper had use of funds, timely recovery of sanctioned loans are some of the problems being faced by primary cooperative agricultural rural development banks. Also, these banks facing a number of challenges such as growing Non-Performing Assets, increased demand for credit and falling deposits. Besides this, these institutions have been facing competition from sister institutions like Cooperatives, Commercial Banks, Development Banks and also from the unorganized sector.

(ii) State Co-Operative Agriculture and Rural Development Banks

These banks operate at state level through their branches in different villages. The State Cooperative Agriculture and Rural Development Bank has been established with a mission to encourage rural farmers to make adequate investment in agriculture by availing the benefit of its loan schemes.

Objectives

The primary objective of the banks is to finance the PCARDBs, who are its members. In order to fulfill this objective, the banks have got the following objectives.

- To float debentures.
- To receive deposits.
- To grant loans to PCARDBs for the purposes approved by NABARD and Registrar of Co-operatives Societies.
- To inspect the PCARDBs and the lands mortgaged to them.
- To develop, assist and coordinate the work of affiliated PCARDBs.

Sources of Funds

The SCARDBs raise the needed finance from the following sources.

- Share Capital
- Issue of Debentures
- Financial assistance from State Bank of India
- Grants and Subsidies
- Deposits and other funds

4.2.2 Commercial Banks

Traditionally, commercial banks are physically located in buildings where customers come to use teller window services, ATMs and safe deposit boxes. But a growing number of commercial banks operate exclusively online, where all transactions with the commercial bank must be made electronically. These "virtual" commercial banks often pay a higher interest rate to their depositors. This is because they usually have lower service and account fees, as they do not have to maintain physical branches and all the ancillary charges that come along with them, such as rent, property taxes and utilities. For many years, commercial banks were kept separate from another type of financial institution called an investment bank. Investment banks provide underwriting services, M&A and corporate reorganization services, and other types of brokerage services for institutional and high-net-worth clients.

A commercial bank is a type of financial institution that accepts deposits; offers checking account services; makes business, personal and mortgage loans; and offers basic financial products like certificates of deposit (CDs) and savings accounts to individuals and small businesses. A commercial bank is where most people do their banking, as opposed to an investment bank.

Commercial banks make money by providing loans and earning interest income from those loans. The types of loans a commercial bank can issue vary and may include mortgages, auto loans, business loans and personal loans. A commercial bank may specialize in just one or a few types of loans.

Customer deposits, such as checking accounts, savings accounts, money market accounts and CDs, provide banks with the capital to make loans. Customers who deposit money into these accounts effectively lend money to the bank and are paid interest. However, the interest rate paid by the bank on money they borrow is less than the rate charged on money they lend.

Commercial Banks (CBs) provide rural credit by establishing their branches in the rural areas. The share of commercial banks in rural credit was very meager till 1969. The All India Rural Credit Review Committee (1969) recommended multi agency approach to the rural and especially agricultural credit. It suggested the increasing role of the CBs in providing agricultural credit. Further, under the Social Control Policy introduced in 1967 and subsequently the nationalization of 14 major CBs in 1969, CBs have been given a special responsibility to set up their advances for agricultural and allied activities in the country. The major expansion of rural branches took place and CBs introduced Lead Bank scheme and district credit plans for rural areas. Banks were asked to lend 18 percent of their total advances to agriculture within the quota of 40 percent of priority sector lending. This expansion of rural credit remained till the late 1980s. However, during late 80's, CBs suffered huge losses due to waiving of agricultural loans by the government. The financial liberalization process with the adoption of Narasimham Committee report in 1993 has necessitated the banks to focus on profitability and adopt prudential norms. The proportion of bank credit to rural areas especially small borrowers has come down steadily.

Customers find commercial bank investments, such as savings accounts and CDs, attractive because they are insured by the Federal Deposit Insurance Corp. (FDIC), and money can be easily withdrawn. However, these investments traditionally pay very low interest rates compared with mutual funds and other investment products. In some cases, commercial bank deposits pay no interest, such as checking account deposits.

Types

Commercial banks are of three types, which are as follows:

(a) Public Sector Banks: Refer to a type of commercial banks that are nationalized by the government of a country. In public sector banks, the major stake is held by the government. In India, public sector banks operate under the guidelines of Reserve Bank of India (RBI), which is the central bank. Some of the Indian public sector banks are State Bank of India (SBI), Corporation Bank, Bank of Baroda, Dena Bank, and Punjab National Bank.

(b) Private Sector Banks: Refer to a kind of commercial banks in which major part of share capital is held by private businesses and individuals. These banks are registered as companies with limited liability. Some of the Indian private sector banks are Vysya Bank, Industrial Credit and Investment Corporation of India (ICICI) Bank, and Housing Development Finance Corporation (HDFC) Bank.

(c) Foreign Banks: Refer to commercial banks that are headquartered in a foreign country, but operate branches in different countries. Some of the foreign banks operating in India are Hong Kong and Shanghai Banking Corporation (HSBC), Citibank, American Express Bank, Standard & Chartered Bank, and Grindlay's Bank. In India, since financial reforms of 1991, there is a rapid increase in the number of foreign banks. Commercial banks mark significant importance in the economic development of a country as well as serving the financial requirements of the general public.

Process

When a commercial bank lends money to a customer, it charges a rate of interest that is higher than what the bank pays its depositors. For example, suppose a customer purchases a five-year CD for $10,000 from a commercial bank at an annual interest rate of 2%. On the same day, another customer receives a five-year auto loan for $10,000 from the same bank at an annual interest rate of 5%. Assuming simple interest, the bank pays the CD customer $1,000 over five years, while it collects $2,500 from the auto loan customer. The $1,500 difference is an example of spread — or net interest income — and it represents revenue for the bank.

In addition to the interest it earns on its loan book, a commercial bank can generate revenue by charging its customers fees for mortgages and other banking services. For instance, some banks elect to charge fees for checking accounts and other banking products. Also, many loan products contain fees in addition to interest charges.

Functions

The functions of commercial banks are discussed as follows:

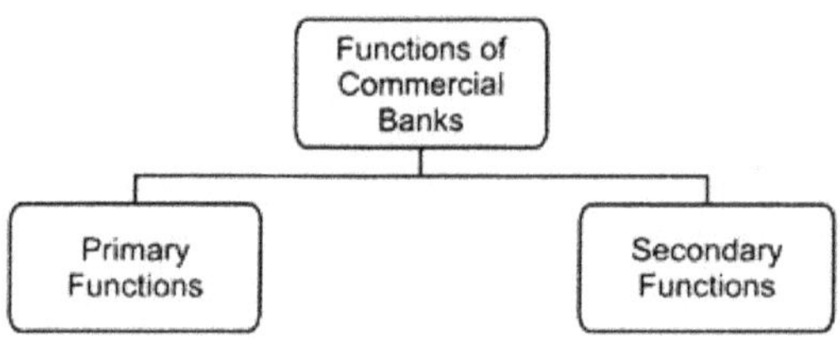

Fig Functions

- Primary Functions: Refer to the basic functions of commercial banks that include the following:
  - Accepting Deposits: Implies that commercial banks are mainly dependent on public deposits. There are two types of deposits, which are discussed as follows:
  - Demand Deposits: Refer to kind of deposits that can be easily withdrawn by individuals without any prior notice to the bank. In other words, the owners of these deposits are allowed to withdraw money anytime by simply writing a check. These deposits are the part of money supply as they are used as a means for the payment of goods and services as well as debts. Receiving these deposits is the main function of commercial banks.

- Time Deposits: Refer to deposits that are for certain period of time. Banks pay higher interest on rime deposits. These deposits can be withdrawn only after a specific time period is completed by providing a written notice to the bank.
- Advancing Loans: Refers to one of the important functions of commercial banks. The public deposits are used by commercial banks for the purpose of granting loans to individuals and businesses. Commercial banks grant loans in the form of overdraft, cash credit, and discounting bills of exchange.
    - Secondary Functions: Refer to crucial functions of commercial banks. The secondary functions can be classified under three heads, namely, agency functions, general utility functions, and other functions. These functions are explained as follows:
- Agency Functions: Implies that commercial banks act as agents of customers by performing various functions, which are as follows:
    - Collecting Checks: Refer to one of the important functions of commercial banks. The banks collect checks and bills of exchange on the behalf of their customers through clearing house facilities provided by the central bank.
    - Collecting Income: Constitute another major function of commercial banks. Commercial banks collect dividends, pension, salaries, rents, and interests on investments on behalf of their customers. A credit voucher is sent to customers for information when any income is collected by the bank.
    - Paying Expenses: Implies that commercial banks make the payments of various obligations of customers, such as telephone bills, insurance premium, school fees, and rents. Similar to credit voucher, a debit voucher is sent to customers for information when expenses are paid by the bank.
- General Utility Functions: Include the following functions:
    - Providing Locker Facilities: Implies that commercial banks provide locker facilities to its customers for safe keeping of jewellery, shares, debentures, and other valuable items. This minimizes the risk of loss due to theft at homes.
    - Issuing Traveler's Checks: Implies that banks issue traveler's checks to individuals for traveling outside the country. Traveler's checks are the safe and easy way to protect money while traveling.
    - Dealing in Foreign Exchange: Implies that commercial banks help in providing foreign exchange to businessmen dealing in exports and imports. However, commercial banks need to take the permission of the central bank for dealing in foreign exchange.

- Transferring Funds: Refers to transferring of funds from one bank to another. Funds are transferred by means of draft, telephonic transfer, and electronic transfer.

- Other Functions: Include the following:
  - Creating Money: Refers to one of the important functions of commercial banks that help in increasing money supply. For instance, a bank lends Rs. 5 lakh to an individual and opens a demand deposit in the name of that individual. Bank makes a credit entry of Rs. 5 lakh in that account. This leads to creation of demand deposits in that account. The point to be noted here is that there is no payment in cash. Thus, without printing additional money, the supply of money is increased.
  - Electronic Banking: Include services, such as debit cards, credit cards, and Internet banking.

Credits Offered

A commercial bank offers short-term loans to individuals and organizations in the form of bank credit, which is a secured loan carrying a certain rate of interest. There are various types of bank credit which are provided by commercial bank such as:

Bank Loan: Bank loan may be defined as the amount of money granted by the bank at a specified rate of interest for a fixed period of time. The commercial bank needs to follow certain guidelines to extend bank loans to a client. For example the bank requires the copy of identity and income proofs of the client and a guarantor to sanction bank loan. The banks grant loan to clients against the security of assets so that, in case of default, they can recover the loan amount. The securities used against the bank loan may be tangible or intangible, such as goodwill, assets, inventory, and documents of title of goods.

Advantages:

- Grants loan at low rate of interest
- Involves very simple process of loan granting
- Requires minimum document and legal formalities to pass the loan
- Involves good customer relationship management
- Consumes less time because of modern techniques and computerization
- Provides door-to-door facilities

Limitations: In addition to advantages, the bank loan suffers from various imitations, which are as follows:

- Imposes heavy penalty and legal action in case of default of loan

- Charges high rate of interest, if the party fails to pay the loan amount in the allotted time
- Adds extra burden on the borrower, who needs to incur cost in preparing legal documents for procuring loans
- Affects the goodwill of the organization, in case of delay in payment

Cash Credit: Cash credit can be defined as an arrangement made by the bank for the clients to withdraw cash exceeding their account limit. The cash credit facility is generally sanctioned for one year but it may extend up to three years in some cases. In case of special request by the client, the time limit can be further extended by the bank. The extension of the allotted time depends on the consent of the bank and past performance of the client. The rate of interest charged by the bank on cash credit depends on the time duration for which the cash has been withdrawn and the amount of cash.

Advantages:

- Involves very less time in the approval of credit
- Involves flexibility as the cash credit can be extended for more time to fulfill the need of the customers.
- Helps in fulfilling the current liabilities of the organization
- Charges interest only on the amount withdrawn by the customer. The interest on cash credit is charged only on the amount of cash withdrawn from the bank, not on the total amount of credit sanctioned.

Limitations:

The cash credit is one of the most important instruments of short-term financing but it has some limitations.

- Requires more security for the approval of cash
- Imposes very high rate of interest
- Depends on the consent of the bank to extend the credit amount and the time limit

Bank Overdraft: Bank overdraft is the quickest means of the short-term financing provided by the bank. It is a facility in which the bank allows the current account holders to overdraw their current accounts by a specified limit. The clients generally avail the bank overdraft facility to meet urgent and emergency requirements. Bank overdraft is the most popular form of borrowing and do not require any written formalities. The bank charges very low rate of interest on bank overdraft up to a certain time.

Advantages:

- Involves no documentation for the extension of overdraft amount
- Imposes nominal interest on the overdraft amount
- Charges fee only on the amount exceeding the account limit

Disadvantages:

- Incurs high cost for the clients, if they fail to pay the amount of overdraft for a longer period of time
- Hampers the reputation of the organization, if it fails to pay the amount of overdraft on time
- Allows the bank to deduct overdraft amount from the customers' accounts without their permission

Discounting of Bill: Discounting of bill is a process of settling the bill of exchange by the bank at a value less than the face value before maturity date. According to Sec. 126 of Negotiable Instruments, "a bill of exchange is an unconditional order in writing addressed by one person to another, signed by the person giving it, requiring the person to whom it is addressed to pay on demand or at fixed or determinable future time a sum certain in money to order or to bearer." The facility of discounting of bill is used by the organizations to meet their immediate need of cash for settling down current liabilities.

Conditions laid down by the bank for discounting of bill are as follows:

- Must be intended to specific purpose
- Must be enclosed with the signature of the two persons (company, bank or reputed person)
- Must be less than the face value
- Must be produced before the maturity period.

4.2.3 Regional Rural Banks (RRBS)

Regional Rural Banks (RRBs) is also known as Gramin banks which are Indian scheduled banks operating at regional level in different states of India. They have been created with a view of serving primarily the rural areas of India with basic banking and financial services. However, RRBs may have branches set up for urban operations and their area of operation may include urban areas too.

The area of operation of RRBs is limited to the area as notified by Government of India covering one or more districts in the State. RRBs also perform a variety of different functions. RRBs perform various functions in following heads:

- Providing banking facilities to rural and semi-urban areas.

- Carrying out government operations like disbursement of wages of MGNREGA workers, distribution of pensions etc.
- Providing Para-Banking facilities like locker facilities, debit and credit cards.
- Small financial banks

RRBs are the specialized banks established under RRB Act, 1976 to cater to the needs of the rural poor. RRBs are set-up as rural-oriented commercial banks with the low cost profile of cooperatives but with the professional discipline and modern outlook of commercial banks. Between 1975 and 1987, 196 RRBs were established with over 14,000 branches. As a result of the amalgamation, the number of RRBs was reduced from 196 to 133 as on 31 March, 2006 and to 96 as on 30 April 2007. RRBs covered 525 out of 605 districts as on 31 March 2006. After amalgamation, RRBs have become quite large covering most parts of the State. Increased coverage of districts by RRBs makes them an important segment of the Rural Financial Institutions (RFI). The branch network of RRBs in the rural area form around 43 per cent of the total rural branches of commercial banks. A large number of branches of RRBs were opened in the un-banked or under-banked areas providing services to the interior and far-flung areas of the country.

Structure

The organizational structure for RRB's varies from branch to branch and depends upon the nature and size of business done by the branch. The Head Office of an RRB normally had three to seven departments. The following is the decision making hierarchy of officials in a Regional Rural Bank.

- Board of Directors
- Chairman & Managing Director
- General Manager
- Chief Manager/Regional Managers
- Senior Manager
- Manager
- Officer / Assist

Role of RRB in Rural Development of Economy

RRBs plays a pivotal role in the economic development of the rural India. The main goal of establishing regional rural banks in India was to provide credit to the rural people who are not economically strong enough, especially the small and marginal farmers, artisans, agricultural labours, and even small entrepreneurs.

RRBs primarily cover small and marginal farmers, landless laborers, rural artisans, small traders and other weaker sections of the rural community. However, even after so many years, the market share of RRBs in rural credit remained low and has suffered huge losses. In recent years Government has initiated reform process to improve the functioning of RRBs.

The role of these banks in economic development is examined, their policies and progress are assessed, and suggestions are given on how they can help improve and uplift the weaker sections of society. This latter objective has been largely achieved by the RRBs. They have succeeded in extending their branch network to the unbanked or under-banked rural areas, and they have made significant progress in deposit mobilization, and in extension of credit particularly among the poor. Furthermore, RRBs have played an important role in the implementation of various special schemes such as the Integrated Rural Development Programme.

In rural areas, RRBs account for a substantial 37% of total offices of all scheduled RBBs. In semi-urban areas, their share comes to 15%. It goes without saying that exclusion is more severe in rural areas 91% of the total workforce in RRBs is posted in rural and semi-urban areas as compared to 38% for other scheduled RBBs . Even in absolute terms out of a total workforce of 179,423 deployed by all scheduled RBBs in rural areas, RRBs share is 25% (45,062). This is significant considering that at all India level, manpower of RRBs constitute only 7% of the total manpower of all scheduled RBBs.

At all India level, RRBs account for 18% of loan accounts of all scheduled RBBs and 3% of loans outstanding. However in rural areas the share of RRBs in loan accounts is an impressive 38%. More significantly, despite having 38%of all loan accounts, RRBs account for only 21% of total credit outstanding in rural areas implying thereby their better reach to small borrowers

The microfinance services provided through self help groups bank linkage has so far been the most successful initiative in financial inclusion. Therefore RRB's involvement in these Self help groups have increased financial inclusion at more than faster rate than it was before

RRBs have not only provided financial services to the SHG-Bank Linkage Program me, but have also played a significant role as self help promotional institution . As many as 104 RRBs (31 March 2006) are also functioning as SHPIs with grant assistance from NABARD.

Problems

Apart from the high cost of their operations, RBBs find it difficult to post the clerical staff to rural centres since the latter prefer to work in urban centers. Even in. respect of transferable cadres like branch ,the banks face a problem in finding the required staff for rural centres .One of the important reasons for the unsatisfactory functioning of many rural branches is the unwillingness of the urban-oriented staff to get involved in rural lending to the fullest extent. Both the clerical staff and supervisory staff who have joined the RBBs with expectations of urban posting and the attendant comforts, are often averse to staying and touring in villages. On balance, the advantage would seem to lie in encouraging RBBs to transfer the eligible of their rural branches to RRBs by them, wherever possible. Rural bank loans to weaker sections may be given priority for such transfer. Similarly, rural branches which have not been able to develop adequate business seven after a reasonable period, so as to cover their own costs of establishment deserve to be considered for this purpose. On the other hand, where the bigger category of borrowers are substantial in numbering any rural branch and account for bulk of the business, it may be retained by the RRBs

The willingness to allow the RRBs these specific considerations notwithstanding, there were serious attempts to redress the prevalent problems of these banks, the principal one at the time being the low recovery rates and loan over dues. These had not only led to capital erosion but more importantly had resulted in non-recycling of funds, which in turn necessitated increasing dependence on external sources of refinancing. The internal factors identified as contributing to low recovery were: weak monitoring and supervision, apathy towards recovery, failure to link lending with development and to ensure end use of the loan. The external factors were: political interference, willful default, drought and floods, underdevelopment, lack of legal and administrative support from the state government in the matter of recovery, etc. Generally, recovery was low in respect of loans sanctioned under IRDP and other poverty alleviation programmes. Besides there were certain organizational problems such as low capital base, multiple ownership and divided responsibility, lack of adequate training for the RRB staff, etc. which called for intervention.

It was estimated that the RRBs would need about seven years to become viable, though for the RRBs with a large number of infant branches even this period might not be adequate. Between 1980 and 1987, while the number of RRBs increased a little more than two-fold, the number of branches of RRBs increased more than four-fold. It was not totally unexpected therefore that by the end of the 1980s several of these banks were showing losses on their books.

4.3 Non-institutional sources of agricultural finance

4.3.1 Introduction

The issues of land clearance and intensification of agriculture cannot be seen as issues related exclusively, or in many cases even primarily, to loan-making decisions by formal banking institutions. While banks are the principal institutional source of capital, they are not always the most important source of credit for agriculture. In fact, in most of the developing world, in terms of aggregate lending volumes, banks generally lend far less to agriculture than does the informal sector.

Therefore, any program designed to confront the problem of the impact of agricultural lending upon the environment must have a broader perspective. The importance of informal capital markets points to the necessity of involving the broader society, and the government in its policy-making role, in the process of protecting the environment and building sustainable systems of agriculture.

To understand the role that banks play in financing agriculture, it is necessary to focus upon the broader capital market and upon the aggregate capital flows from all sources to the agricultural sector . There are three salient sets of features that characterize rural credit markets in the developing world.

First, banks, especially those in the developing countries, lend very little of their total resources outside urban areas and even less to agriculture. Their presence in the rural areas is more frequently in order to capture low cost deposits than to carry out an extensive lending program.

Second, in most developing countries, the informal market is a far more important source of capital than are formal banking institutions which supply a relatively small part of the total volume of credit flows to agriculture. Most production credit, as well as a fair share of term lending, comes not from formal banking institutions but, instead, from the informal market.

In most countries, banks offer lower nominal lending rates to farmers than do informal lenders. However, the processing time, repeated travel from the borrower's home to the bank, the paperwork, literacy requirements, the demanding loan conditions as to equity, collateral, and credit histories, developed by bureaucrats in a distant urban area, as we]1 as the need to repay the loan in cash, add substantially to the cost of borrowing for small farmers. As a result, the effective borrowing rate, even in subsidized credit programs, is quite high. So high, in fact, that many farmers who could borrow at a formal banking institution opt for borrowing in the informal market.

Third, government seldom has adequate resources to meet more than a small portion of the total demand for credit. Total demand for agricultural production credit is estimated at about 60 billion. About 24 billion is supplied by the private banks (including commercial, thrift and rural banks). The government controlled only about one billion of agricultural production credit, or about 2 % of the demand.

4.3.2 Sources

The non-institutional finance forms an important source of rural credit in India, constituting around 40 percent of total credit in India. The interest charged by the non-institutional lenders is usually very high. The land or other assets are kept as collateral. The important sources of non-institutional credit are as follows:

(a) Money-Lenders

Money-lending has been the widely prevalent profession in the rural areas. The moneylenders charge huge rate of interest and mortgage the property of the cultivators and in some cases even the peasants and members of his family are kept as collateral.

Moneylenders have been around for generations, but their business has boomed ever since India's economic priorities shifted, with globalization, from agriculture to industry. The arrival of high-cost seeds and pesticides has added to the debts. Almost every farmer in India's massive rural swathes is tethered, in one way or another, to the sahukar, the Indian variety of the moneylender, the ubiquitous, ravenous loan shark.

For centuries, moneylenders have monopolized rural Indian credit markets. Families have lost land, farmers have been asked to prostitute their wives to pay off debts, and, when all else has failed, they have tied the noose to end their misery.

An inescapable cycle of debt continues to grip rural India, particularly its farming class. Yet the public image of menacing debt collectors does not reflect the actual plight of India's three million farmers. Moneylenders have been around for generations, but their business has boomed ever since India's economic priorities shifted, with globalization, from agriculture to industry. The arrival of high-cost seeds and pesticides and the attraction of bumper harvests have added to the debts. In farm belts moneylenders operate under the guise of farm input sellers.

Moneylenders are a key part of India's economy. They charge higher, sometimes usurious, interest rates but require few formal guarantees and offer hassle-free services. The nation has just 13 bank branches per 100,000 people and only about one in four people have access to the internet. Single households run about 96 percent of the nation's unincorporated non-farm enterprises and only 1 percent received loans from the government. For the bulk of low-income households, moneylenders are the only dependable source for money when emergencies arise.

Almost every farmer in India's massive rural swathes is tethered, in one way or another, to the 'sahukar'. He is the Indian variety of the moneylender, the ubiquitous, ravenous loan shark. For centuries, moneylenders have monopolised rural Indian credit markets. Families have lost land and farmers have been asked to prostitute their wives to pay off debts. And, when all else has failed, they have tied the noose to end their misery.

An inescapable cycle of debt continues to grip rural India, particularly its farming class. Yet the public image of menacing debt collectors does not reflect the actual plight of India's three million farmers. Moneylenders have been around for generations. But their business has boomed ever since India's economic priorities shifted, with globalisation, from agriculture to industry. The arrival of high-cost seeds and pesticides and the attraction of bumper harvests have added to the debts. In farm belts, moneylenders operate under the guise of farm input sellers.

In Maharashtra, farmers' dependence on private money lenders has shown a steep rise at 40 per cent in the last one year The total amount of loans disbursed by private money lenders rates was Rs 1,254,97 crore. Compared to the 2015 statistics, the total amount disbursed thus has risen by Rs 358.63 crore.

According to the All India Debt Investment Survey of 2012, nearly 48% farmers across the country took loans from informal sources such as from moneylenders and landlords. The number had risen from 36% in 1991 and 43% in 2001. Among farmers who owned land parcels smaller than 0.1 hectares, 85% had pending loans from such informal finance sources.

Farming distress has attracted a new breed of moneylenders. Anyone with some disposable cash, from shopkeepers, government officials, and policemen to village teachers, can now lend in the hope of making a killing. They are willing to extend credit but at highly extortionate rates (sometimes exceeding 50 percent). This keeps the borrowers in lifelong penury.

While these small farmers pay exorbitant interest, the affluent farmers get subsidised credit. The government's interest subvention (subsidy) scheme for farmers provides credit at a subsidised interest rate of 7 per cent and for prompt re-payers at 4 per cent.

It was expected that in socialist India, banks would become an extremely popular port of call for clients seeking loans. In fact, these financial institutions recorded a surge in the social banking era of the 1970s but the populist policies left a cruel legacy of dud loans. This sour experience made bankers very wary and they turned off the spigots. Institutional credit is now mired in thickets of red tape blocked by bankers who are bedevilled by a highly contaminated credit culture. Hence, moneylenders continue to thrive.

A current of dread runs through the country's suicide-ravaged farmlands as their debts pass from husband to widow, from a father to his children. Most villages are locked into a bond with village moneylenders — an intimate bond, and sometimes a menacing one. Popular cinema and classic literature tell many pathos-filled narratives of India's poor caught in that karmic cycle of poverty. Those stories inevitably end in tragedy.

Farmers who fall into the trap of money lending find themselves locked in a white-knuckle gamble. They juggle large loans at uncertain interest rates, in the hope that someday a bumper harvest will allow them to clear their debts — so they can take out new loans. This pattern has left a trail of human wreckage.

There are two types of money lenders in rural areas. There are rich farmers or landlords who combine farming with money-lending. There are also professional money lenders whose only occupation or profession is to lend money. The cultivators depend upon the money-lenders for their requirements of cash. However, there are many reasons for the preponderance of the village money-lenders in rural area even now.

- ❖ The money lender freely supplies credit for productive and non-productive propose, and also for short-term and long-term requirements the farmers.
- ❖ He is easily accessible and maintains a close and personal contact with the borrowers often having relations with family extending over generations.
- ❖ Their methods of business are simple and elastic.

(b) Traders, landlords and commission agents

The agents give credit on the hypothecation of crops which when harvested is used to repay loans.

Traders

Traders supply funds to farmers for productive purpose much before the crops mature. They force the framers to sell their produce at low price and they charge a heavy commission for themselves. Thus source of finance is particularly important in the case of cash crop like cotton, groundnut, tobacco, and in the case of fruit of chard like mangoes. Traders may be bracketed with money lenders, as their lending to farmers is also at exorbitant rates and has other undesirable effects too.

Commission Agent

Commission agents--also known as commercial agents--work as middlemen between vendors and buyers. These individuals find employment in myriad industries for large and small businesses, depending on their area of specialization. Commission agents work in countries throughout the world and may seek work from a number of employers simultaneously if their field of specialization provides the opportunity to do so. These expert purchasers and vendors provide a valuable service to their clients.

In general, commission agents purchase and sell items on behalf of a principal, usually a company. The nature of the job depends on the field of employment. Agriculture produce commission agents who work for small businesses, for instance, travel to farms and orchards, purchasing fruits, vegetables and dairy products for principals such as grocers and restaurants, or sell such items to grocers and restaurants on behalf of farmers. These individuals work independently as contracted, third-party workers, not employees of their principals. Commission agents conduct business under their own names, which affords a measure of anonymity to their principals while allowing the agents a certain degree of autonomy.

As experts in their field, commission agents can assist small businesses in making the most of their limited budgets.

Commission agents possess a fair amount of freedom when it comes to purchasing or selling materials for a principal. Usually, principals outline their material needs and a project's purchase budget, or the desired price of sale, along with a set of preferred conditions for a deal. The commission agent is responsible for meeting the ideal conditions requested by the principal, and must stay within the provided budget or the price of sale. As long as commission agents conform to these stipulations, they possess the freedom to act and make deals as they choose.

(c) Credit from relatives

Credits are generally used by farmers for meeting personal expenditure. Calculating how much is spent on family living expenses is never easy. But it can be especially hard when the farmer and the business are tied closely together, as they are on the farm. To control family finances, farmer have to know where the money is coming from and where it goes.

As farmer begin to examine their family living expenses, he/ has to consider these aspects. Farm income can be irregular and uncertain. Income often varies considerably from year to year. Income and expenses usually follow a seasonal pattern, but even that pattern varies from year to year.

Family expenditures, such as the mortgage and utility bills, often seem lower than for non-farm families. However, that can be deceptive because these costs are often included with farm-business expenses rather than family living expenses. For example, the mortgage payment may include financing for both the house and the land, but the amounts for each may not be listed separately.

Food expenditures may be lower than for non-farm families. Homegrown meat and produce, for example, may trim grocery bills. But, for families with many in-town activities, food eaten away from home can add greatly to food expenses.

Insurance and health care costs can be high. There may not be an employer to pick up part of the cost of employee insurance plans. The farm family may not participate in a group insurance plan. And the need for insurance is magnified because agricultural producers face a greater risk of injury, disability, or death. There is competition between your family and your farm for the use of cash. It is often difficult to decide whether to invest farm-business profits back into the farm or to make expenditures for the home or the family.

(d) Brokers

Broker is a matching agent who arranges a transaction between a buyer and a seller, and for that service, is paid a commission. The seller typically pays a commission, but some brokers may charge commission to both the seller and buyer.

A cash grain broker acts as a matching agent between a seller and a buyer of physical grain. As a broker, they do not take legal or physical possession of the crop being brokered, but arrange for the deal to proceed upon agreement of the terms of the transaction by both parties. The broker may arrange for the trucking, or the buyer or seller may have the trucking capacity to handle that aspect. Unlike a grain dealer, a cash grain broker does not take legal possession of the crop and therefore does not need to be licensed, although some brokers may elect to do so. If licensed, the broker will need to post a bond or irrevocable letter of credit from a bank to serve as security for outstanding payables.

Cash grain brokers often deal with buyers who are not licensed. Under current rules, end users of grain need not be licensed. Crop producer protection has been under review in recent years and is subject to change. End users obviously include cattle, hog and poultry feeding operations. The cash grain broker attempts to deal with buyers who they believe will stand for payment of the delivered grain, but there is always a chance of a full or partial payment default. In such a case, the cash grain broker will try to help recover the funds. At least one cash grain broker, in a case of payment default, covered the outstanding payments for their brokered grain sales from their own resources.

4.4 Commercial banks as a source of agricultural finance

Social control of commercial banks

The proposed Bill on social control of banks will provide for reconstitution of the Boards of Directors and prohibit advances or guarantees to Directors and the concerns in which they are interested. The Bill will also provide for acquiring the business of a particular bank if it defaulted in complying with the provisions of law or directives issued thereunder. The Bill, which will confer additional powers on the Reserve Bank, will be introduced in the next few days. In a statement in the Lok Sabha to-day [December 14, New Delhi] announcing the decision of the Government to introduce social control over commercial banks, the Deputy Prime Minister and Finance Minister, Mr. Morarji Desai, stated emphatically that mere acquisition of banks would only severely strain the administrative resources of the Government while leaving the basic issues regarding banking system untouched.

The growth of commercial banking during the first three plan periods has been lopsided. Without demanding proper security some banks were diverting funds to large and medium industries. Bank branches were opened only in big cities. Rural areas were neglected. Banks made discrimination between private sector and public sector between rural and urban and between agriculture, trade and industry. The banks were financing those industries which were producing luxury goods. The government felt that this type of growth was not in consonance with planning. The growth appeared to be defiant in many respects. There was a growing demand for the bank credit for the development of agriculture, industry and self employment.

So it was essential for the banking system to attract savings. Therefore, the government felt that need for social banking as against capitalist banking. A scheme of social control was introduced in 1967. The government enacted Banking Laws Amendment Act in 1968.

This act has given more power to the government to control banking. The objectives of this Act was to ensure more equitable distribution of the resources of the banking system. The priority sectors like agriculture, small-scale industry, public sector and self-employment were to receive their due share in obtaining bank finance. Apart from this the banks are required to reconstitute their board of directors.

The government set up National credit council in 1968. The Finance Minister was the chairman and the Governor of the Reserve Bank was the vice chairman of the council.

The main functions of the National credit council were:

- assessing the volume of credit required for the economy as a whole.
- providing guidelines for the distribution of credit to the priority sector.
- ensuring equitable distribution of credit in the economy.

4.5 Nationalization of commercial banks

4.5.1 Introduction

Nationalization is a process whereby a national government or State takes over the private industry, organisation or assets into public ownership by an Act or ordinance or some other kind of orders. This strategy has been frequently adopted by socialist governments for transition from capitalism to socialism.

The banking sector in India has been facing extreme changes with the economic growth of the country. In 1948, RBI (Transfer of public ownership) Act was passed to nationalised the Reserve Bank. On Jan 1, 1949, RBI was nationalised. In 1955, the Imperial Bank of India was nationalized and was given the name "State Bank of India", to act as the principal agent of RBI and to handle banking transactions all over the country. It was established under State Bank of India Act, 1955.

These seven associate banks are – the State Bank of Hyderabad, the State Bank of Jaipur and Bikaner, the State Bank of Travancore, the State Bank of Mysore, the State Bank of Patiala, the State Bank of Indore, and the State Bank of Saurashtra.

The third major step was the nationalisation of 14 major commercial banks with deposits exceeding Rs. 50 crores each on 19 July 1969. They are: Allahabad Bank, Bank of Baroda, Bank of India, Bank of Maharashtra, Canara Bank, Central Bank of India, Dena Bank, Indian Bank, Indian Overseas Bank, Punjab National Bank,

Syndicate Bank, Union Bank of India, United Bank of India, and United Commercial Bank.

The last step was the nationalisation of 6 more commercial banks on 15 April 1980. They were: Andhra Bank, Corporation Bank, New Bank of India, Oriental Bank of Commerce, Punjab and Sind Bank, and Vijaya Bank. At the time of nationalisation, each one of these had crossed the deposit mark of Rs.200 crores, though at the time of nationalisation of 14 banks their individual deposits were below Rs.50 crores.

In June 1996, there were 19 nationalised banks, New Bank of India was merged with Punjab National Bank in 1993. We discuss below the justification, objectives, achievements and failures of nationalisation of commercial banks in India. As nationalisation of major commercial banks is an established fact now, the case against nationalisation is uncalled for and hence of purely academic interest.

4.5.2 Justification

The nationalisation of the commercial banks in India was a very controversial issue till the nationalisation of 14 major banks in July, 1969. A number of arguments were advanced by economists, political leaders, bankers, workers, and the public in support of nationalisation of banks which are enumerated below:

1. Neglect of Priority Sectors: A major charge against banks in India was that they advanced loans to large scale industries, and neglected the priority sectors such as agriculture, small scale industries, etc. The banks advanced 60% of the total bank credit to big industries and 19% to trade in March 1968. On the other hand, the share of agriculture was 2.2%, of small industries 6.9%, and of retail trade less than 2% in the total bank credit. It would be possible to direct bank credit on a large scale through bank nationalisation.

2. Neglect of Plan Priorities: Another charge was that commercial banks neglected plan priorities and diverted credit into less important channels. Bank nationalisation would help the Government in directing bank credit in accordance with plan priorities.

3. High Profits: It was alleged, that banks earned huge profits which were disproportionate to the capital invested in the banks. The ratio of profits of scheduled banks before tax to the paid-up capital and reserve increased from 12.2% to 42.5% in 1968. In other words, they earned Rs.42.5crores annually. On the other hand, their capital base was declining. The ratio of paid-up capital and reserves to deposits fell from 9.7% in 1951 to 2.6% in 1968.

This was in contravention of RBI's directives to banks to keep at least 6% of their deposit liabilities in reserve. Whereas the banks earned huge profits and used them for their interests, the shareholders earned almost nothing because they had hardly more than 1% of deposits as share capital. This was one of the arguments advanced by the Government for bank nationalisation.

4. Funds for Plans: By nationalising the banks, these huge profits of the banks would be available to the Government for financing plans and achieving the plan objectives.

5. Concentration of Economic Power: It was alleged that commercial banks had led to the concentration of wealth and economic power in the hands of a few big business houses and industrialists. For instance, the Central Bank of India Ltd. was being controlled by the Tatas, the Bank of India Ltd. by the Mafatlals, the Punjab National Bank Ltd. by the Dalmias, the United Commercial Bank Ltd. by the Birlas, etc. These big business houses, in turn, controlled a number of industries, thereby enjoying a virtual monopoly in their business.

Moreover, about 90% members of the Board of Directors of banks consisted of industrialists and big businessmen. A single director of a bank was a common director of many other industrial concerns and with his control over the bank he was able to turn the bank for the benefit of his industrial house.

This interlocking of directorship in banks led to unusually high amount of credit to big business houses. Bank nationalisation was, therefore, essential to stop the misuse of the resources of the banks for the benefit of directors and their concerns, and to reduce the concentration of economic power and wealth in a few hands.

6. Socialist Arguments: It was argued that bank nationalisation, was essential for the attainment of a socialist society in India. Such a step would not only reduce concentration of economic power and wealth but also give the State, control over the resources of the banks to be utilised for the masses.

7. Regional Imbalances: It was alleged that commercial banks were increasing regional imbalances. Banking was concentrated in big cities, urban centres and a few States. The Gadgil Study Group of the National Credit Council (1969) vividly brought out the imbalances in banking in the country.

It pointed out that till December 1967, out of 5,64,000 villages in India, not even 1% were served by commercial banks. In 13 districts of the country, there was not a single commercial bank office. More than 600 towns did not have commercial banking facilities.

There were 63 districts in which per capita credit was less than one rupee, in another 13 districts, credit was not extended even to a single individual, although deposits were mobilised from there. The two metropolitan cities of Mumbai and Kolkata accounted for one-third of total deposits and one-half of total credit disbursed by banks.

The ratio of credit to deposits was more than 100% in the metropolitan cities, which means that deposits mobilised in smaller centres were diverted to these cities for financing the activities of big business. It was argued that bank nationalisation would remove these imbalances, ensure even distribution of banking facilities, and spread banking to rural unbanked and under-banked areas.

8. Ignored RBI Directives: Big banks with strong financial background often ign6red the Reserve Bank's directives under the Banking Regulation Act, 1949. This, in turn, adversely affected the RBI's monetary policy. By nationalising such banks, the RBI would be in an advantageous position to implement its monetary policy more effectively.

9. Protection to Depositors: It was argued that bank nationalisation would remove the fear of bank failures from the mind of public, thereby protecting the interests of depositors. This, in turn, would inculcate banking habits among the people, especially in the rural non-monetized sector of the economy.

10. Check of Undesirable Activities: The commercial banks often indulged in anti-social and undesirable activities. They advanced credit for such activities which encouraged speculation, profiteering black marketing etc. which led to accumulation of black money in bank accounts. All such activities would be stopped when banks were nationalised.

11. Encouraged Inflation: It was alleged that commercial banks encouraged inflation by increasing the supply of money without due regard to the directives of the RBI. Banks often used their resources to the discretion of their directors for their own interest and that of their business friends. So they did not follow the guidelines issued by the RBI for credit restriction. This led to inflation. Thus nationalisation of banks was considered to be the only remedy.

12. Increase in Efficiency: It was argued that bank nationalisation would improve the efficiency of banks and the quality of service to the customers because the employees would have a greater sense of security. The efficiency of banks would also increase due to the standardisation of banking operations throughout the country.

13. Better Management: With the nationalisation of banks, the Government would be in a position to develop and provide adequate training and modern managerial techniques to the staff. This would lead to greater efficiency in bank operations.

14. Better Service Conditions: It was hoped that bank nationalisation would provide better service conditions for the employees. They would have the satisfaction of being Government employees. This would also improve employer-employee relations and pave the way for a smooth working of the banking system in the country.

15. Success in Other Countries: The protagonists of bank nationalisation argued that in a number of developed and developing countries where commercial banking was in the hands of State, it had been highly successful in helping the masses and mobilising resources for development.

4.5.3 Objectives

Bank have been nationalised for fulfilling various socio-economic objectives such as:

- ❖ To mobilise savings of the people to the maximum possible extent and utilise them for productive purposes;
- ❖ To ensure prompt operations of the banking system for a larger social purpose and subject it to close public regulation;
- ❖ To meet the legitimate credit needs of private sector industry and trade (big or small);
- ❖ To ensure that the needs of the productive sectors of the economy and, in particular, those of farmers, small scale industrialists and self-employed professional groups are met in an increasing manner.
- ❖ To instruct the banks to provide banking facilities to the hitherto neglected and backward areas in different parts of the country; and
- ❖ To check (stop) the use of the bank credit for speculative and other unproductive purposes.

4.5.4 Performance

The philosophy of bank nationalisation was that those financial institutions which mobilised saving of the public should broadly function as an instrument for promoting economic and social development in more purposive manner. In the post-nationalisation period, there has been a rapid growth of India's banking system which covers:

1. Deposit Mobilisation: There has occurred a significant increase in deposits of scheduled commercial banks in the post-nationalisation period. At the end of June 1969, deposits of these banks were Rs. 4,564 crores. By March 2001, total deposits increased to Rs. 983,268 crores.

It may be noted that deposits mobilised by banks are utilised for two purposes:

- ❖ investments on Government securities and other approved securities in order to fulfill the statutory liquidity requirement (which is 25% at present) and
- ❖ loans and advances to borrowers.

2. Branch Expansion: As against 8,262 branches at the end of June 1969, the total number of commercial bank branches at the end of March 2001 was 63,380. As a result of this, banking coverage in the country as a whole has been improved from one office for 65,000 persons to 15,000 persons during the same period.

3. Coverage of Rural Areas: In the post-nationalisation period, the thrust of the branch expansion policy of commercial banks has been on improving the availability of banking facilities in rural areas. The number of rural branches increased from 1,860 in 1969 to 32,890 in 1997.

4. Credit Deployment: Advances in whatever form constitute the end objective or purpose of banking. From a modest Rs. 3,599 crores in June 1969, total advances' by public sector banks increased to Rs. 265,554 crores in March 1999.

5. Sectoral Allocation: More significant than the increase in bank credit are the changes in sectoral development. In the pre-nationalisation period, large and medium industries as also wholesale trade accounted for more than 79% of total commercial bank credit.

By March 1999, the share of these sectors (including credit for public food procurement) had declined to about 21%; correspondingly, the share of priority sectors and food procurement agencies had shown a significant increase. In recent year's food credit by commercial bank increased substantially because of large volumes of procurement and stock of food-grains. Non-food credit fell reflecting a slowdown in industrial activity.

6. Advances to Priority Sectors: The expansion of credit to small borrowers in the hitherto neglected sectors of the economy has been one of the major tasks of public sector banks in the post- nationalisation period. To achieve this objective, banks have drawn up schemes to extend credit to small borrower in sectors like agriculture, small scale industry, road and water transport operators, retail trade and small business, which traditionally had very little share in credit extended by banks.

Taking into account the need to meet resource requirements of weaker sections, for specific purposes, consumption credit (with certain limits) had been incorporated in priority sector advances. Similarly, small housing loans (not exceeding Rs. 5,000) to the weaker section of society (such as SCs and STs) are also classified as priority sector advances.

Total outstanding credit by banks to small scale industries increased from Rs. 810 crores in June 1969 Rs. 42,591 crores in March 1999. Outstanding to road and water transport operators stood at Rs. 3,620 crores in March 1999.

7. Credit to Weaker Sections of Society: To increase the flow of bank credit to poorer sections comprising small and marginal farmers, landless labourers, tenant farmers and share-croppers, artisans, village and cottage industries and small transport operators, several new credit schemes have been evolved. This section received very little bank credit before nationalisation. In March 1999 the out-standings to small businesses were Rs. 4,231 crores, professional and self-employed person 2,630 crores, housing Rs. 5,366 crores, and consumers and others Rs. 1,108 crores.

8. Direct Finance to Agriculture: Public sector banks were initially given a target of extending 15% of total advances as direct finance to agriculture, to be achieved by March 1985. As against this, advances by public sector banks to priority sectors rose to 16.8% of their total advances by March 1988. Direct finance to agriculture (outstanding) increased from Rs. 310 crores in June 1969 to Rs. 31,167 crores in March 1999. Indirect finance (outstanding) stood at Rs. 6,464 crores.

4.5.5 Benefits and Failure

Benefits

1. Branch Expansion: There has been a sea-change in the banking landscape, both geographically and functionally. The objective of creating an infrastructure for banking activity has, by and large, been achieved. From 6,596 branches and offices in June 1969, their number increased to 32,643 in June 2003.

2. Deposit Growth: Besides, there has been functional diversification in the form of growth in bank deposits. The deposits of these banks in June 1969 were Rs..3,897 crores which increased to Rs.6,88,361 crores in March 2003 and grew by 11.47 per cent during 2002-03 over 2001-02.

3. Credit Expansion: There has been an impressive record of the nationalised banks in the realm of credit expansion. The amount advanced by them increased from Rs.3,035 crores in June 1969 to Rs.3,60,147 crores in March, 2003 and grew by 14.12 per cent during 2002-03 over 2001-02.

4. Priority Sector Lending: These banks have undergone a major transformation from 'wholesale' banking to 'retail' banking. In other words, there has been greater emphasis on disbursing credit to the priority sector comprising' agriculture, small scale industry, transport operators, retail trade, professional and self-employed persons, and education.

The priority sector lending by the nationalised banks increased from Rs.441 crores in June 1969 to Rs.2,03,095 crores in March 2003. As percentage to net bank credit, the same increased from 14.6% to 42.5 during the above period. Agricultural advances increased from 5.4% to 15.3%, the share of credit to SSI units increased from 8.5% to 11.1 and the share of other priority sectors in overall net bank credit increased from 0.7% to 15% during the same period.

5. Spread of Banking in Rural Areas: Since nationalisation, the spread of banking has increased considerably in rural areas. As a result of the deliberate policy of opening a large number of branches in rural areas, 63.1 % of the branches of the 19 nationalised banks were operating in rural and semi-urban areas in June 2003 as against 22% on the eve of nationalisation in June 1969.

6. Reduction in Regional Disparities: Regional imbalances in the availability of banking facilities have been considerably reduced. The majority of these branches have been opened in unbanked areas. In June 1969, there were 74 bank branches in Assam, 172 in Haryana, 42 in Himachal Pradesh and only 2 in Nagaland.

In June, 2003 their number had increased to 1,212; 1,580; 784 and 70 respectively. 35% of branches had been opened in those States where banking facilities were less. Consequently, the average population per bank office declined to 16,000 in June 2003 against 65,000 in June 1969.

7. Some New Schemes: Public sector banks have been departing from their traditional role and introducing some new schemes. Besides launching a range of new savings and other innovative schemes, these banks have taken up merchant banking and leasing and have also been floating mutual funds and venture capital funds.' A number of other steps have been taken up to improve customer service. Courier services have been introduced to expedite collection of outstation credit instruments.

Nomination facilities have been extended to the depositors to avoid hardship to the legal heirs of deceased account holders. Customer service centres have been set up in all major cities. Customer relations programmes are organised periodically.

The system of personal hearing of the grievances of customers has been introduced at appropriate levels to serve the customer better. Nine nationalised banks have opened offices in foreign countries thereby opening new vistas in international trade and finance.

Failures

Despite the above enumerated achievements, the public sector banks suffer from a number of weaknesses which are discussed as under:

1. Deterioration in Customer Service: One of the important drawbacks of nationalisation of banks has been deteriorating customer service. The evils connected with public sector undertakings have crept in these banks thereby resulting in red-tapism, inordinate delays and lack of responsibility.

Despite the directives of the RBI for improving the quality of customer service, little has been done to implement it at the branch level. For instance, there are delays in opening of the counters and sometimes counters are closed for short periods on flimsy grounds causing inconvenience to customers.

The branch managers are helpless to control the staff due to the spread of unionism. Complaint boxes are not maintained in many branches nor are the addresses of regional/zonal managers displayed to help customers who want to appeal to higher officers.

Facilities like automatic credit of outstation cheques of Rs.5000 and above and crediting of interest at savings bank rate in case of collection of outstation cheques being delayed beyond three weeks are not being extended to customers. Moreover, most bank branches give demand drafts against cash to businessmen towards the end of the day.

2. Political Interference: There has been much political interference in granting of loans by public sector banks which has destroyed the sanctity of banking norms. This defeats the very purpose of social banking and leads to malpractices. An important instance is politically organised loan melas.

3. Social Banking Inadequate: Social banking in the form of flow of credit to the priority sector is inadequate. It is less than the total advances of public sector banks. Even today these banks mainly cater to the needs of the corporate sector.

4. Window-dressing of Balance Sheets: Some of the public sector banks have been resorting to window- dressing of their balance sheets so that their balance sheets do not reveal the actual state of affairs. They artificially step up their deposits in the last week of the fiscal year. This has been possible because auditing in the banks is incomplete and partial.

5. Large Scale Irregularities: A number of public sector banks have been resorting to large scale irregularities in their operations which covers:

- ❖ Indulging in Share Speculation: A few banks have started Mutual Funds and subsidiary financial companies and instances have been found of their involvement in share speculation.
- ❖ Favouritism in Granting Loans: Some of the banks are guilty of favouring certain companies in granting loans.
- ❖ Bad Quality of Loan Portfolios: Despite the bad quality of loan portfolios, the banks have been advancing loans under extraneous pressures. Some of the banks have been extending letters of credit and guarantee limits as a routine affair without adopting normal commercial procedure in granting loans.

The non-performing assets (NPAs) of banks reflect the quality of their loan portfolios. NPAs include sub-standard assets, doubtful assets, loss assets, etc. During 1999-2000, the gross NPAs of public sector banks were Rs.53,294 crores, constituting 14.0 per cent of their gross advances.

- ❖ Irregularities in Accounts: Some of the public sector banks have been resorting to large scale irregularities in operating their accounts. In certain cases, .bad accounts have been taken over by one bank from the other without proper scrutiny and judgment.

- Large Bad Debts: These banks have accumulated large amounts of bad debts. As on 31 December 1999, there were 3,185 cases of frauds involving Rs.640 crores of public sector banks.

6. Declining Profitability: The most important problem of the nationalised banks has been their declining profitability. Irregularities and corruption in lending operations, misappropriation, fraud, rising operating cost, deterioration in customer services, etc. have led to declining profitability of these banks. The ratio of net profit to total assets was (-) 1.15% in 1993-94 which improved to only 0.96 per cent in 2002-05.

7. Some Banks as Sick Units: In October 1990, the Reserve Bank of India in a survey found all the nationalised banks in a bad shape and identified eight banks as 'sick'. The 'sick' included: New Bank of India, Punjab and Sind Bank, Bank of India, UCO Bank, Vijaya Bank of India, Syndicate Bank, Bank of Maharashtra, and United Bank of India.

Some of these banks had eroded their reserves and paid-up capital and there had been erosion of their deposit base. To this 'sick-list' were added: Central Bank, Indian Overseas Bank, Allahabad Bank, Andhra Bank, and Dena Bank, in 1992.93.

New Bank of India became so sick that the Government merged it with Punjab National Bank effective 4 September 1993. During 1999-2000, three nationalised banks had been identified as weak. To adjust the losses of weak banks, the Government had contributed Rs. 20,466 crores towards their recapitalisation by 31 March 1999. But no amount was provided during 1999-2000. However, it provided Rs.297 crores towards writing down of the investments of Vijaya Bank for adjustment of its losses.

8. Low Capital Adequacy Ratio: Every bank is required to achieve minimum capital to risk assets ratio, known as capital adequacy ratio (CAR) of at least 9 percent. Out of 27 public sector banks, two banks had not achieved it during 2002-03.

5

Micro Financing in Agriculture

5.1 Microfinance Contributions to Agriculture in Developing Countries

Agriculture has returned to being a primary global concern. Prices of agricultural goods, such as cereal and milk, are hitting record highs on international markets. Food security is no longer guaranteed. High food prices are a short-term problem for consumers, especially among poor population in southern countries. On the other hand, high-performing agricultural markets have the potential to stimulate a new wave of investment and innovation. They give struggling peasants in southern countries a chance to earn a decent living. But to take advantage of this historical situation that so many have been waiting for, farmers will need to invest and increase production.

Access to finance is therefore decisive. Yet the majority of peasants in developing countries are still excluded from the banking system. Bank penetration rates in agricultural regions of Africa and South Asia are barely over 5-6%. In developed countries, agricultural banks played an important role in modernization of agriculture and bankerization early on. In developing countries, microfinance has the potential to play this role, given its advantage in terms of proximity to the client and its frequent association with cooperative approaches. These characteristics are can be found in the approaches originally used by local agricultural banks in France.

5.1.1 Agricultural Finance: Insufficient and Poor Supply Adaptation

The old rural finance paradigm of the 1960s and 1970s was based on public authorities' desire to facilitate access to rural finance. The objective was to promote agricultural development by modernizing agriculture. The most common approach involved direct government intervention via state-owned development banks and direct donor intervention in credit markets with favorable terms and conditions like soft interest rates or lenient guarantees. However, this system was costly and unsustainable, due to poor repayment, and ultimately did not have the desired effect on the development of agriculture production.

In the 1980's, the failure of government-led credit supply gave way to a new paradigm and a renewed approach to rural and agricultural finance in developing countries. State-owned development banks closed, financial sectors were liberalized and microfinance evolved. Based on an approach that encourages financial intermediation, the new rural finance paradigm redefined the roles of the various actors involved in providing financial services, especially governments. Public subsidies were redirected towards creating new microfinance institutions (MFIs) that were supposed to achieve financial sustainability thanks to cost-covering interest rates.

Despite the great hopes associated with the strong growth of the microfinance sector, it soon became clear that the supply of microfinance for agricultural activities was marginal at best and poorly adapted. At the same time, with the liberalization of the financial sector, commercial banks did not pick up the slack of former government-led interventions in rural areas; many banks actually closed their rural branches. Finance for agricultural activities became even more rare.

5.1.2 Microfinance's limited contribution to agricultural finance

Yet despite the important contribution of agriculture to the GDP of the poorest developing countries, the supply of financial services to farmers is still limited. As figure illustrates, the more rural populations contribute to GDP and the greater the percentage of agricultural workers, the lower the rate of financial inclusion.

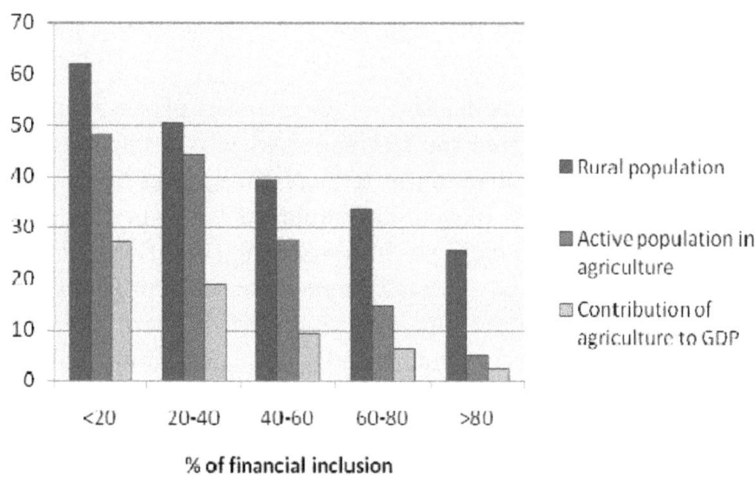

Fig: Contribution of the agricultural sector to GDP and financial inclusion

A regional analysis conducted by Trivelli and Venero of 2004 data from 16 Central and South American countries illustrates this situation. With few exceptions, the portion of agricultural loans disbursed was inferior to the overall contribution of agriculture to the country's GDP.

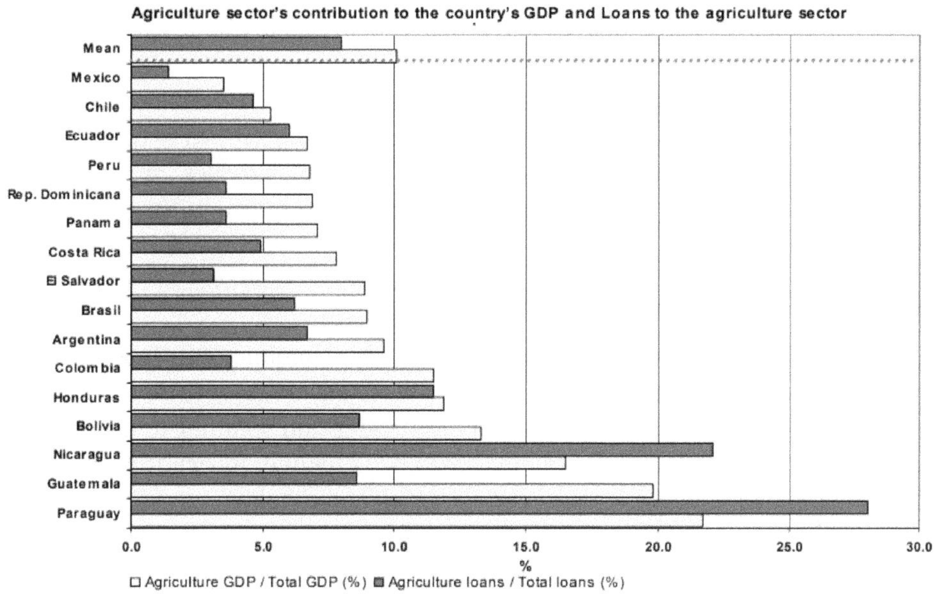

Fig: Portion of credit allocated to agriculture compared to overall contribution of agriculture to GDP

It is very difficult to evaluate the volume of credit microfinance institution's disburse for agriculture. Cumulatively, MFIs around the world lend roughly 30 billion dollars, but there are huge territorial differences in how these funds are allocated.

Indeed, despite annual growth of 36% in the number of poor clients reached by MFIs between 1997 and 2005, most institutions still concentrate on urban and peri-urban areas that are easy to serve. Among rural institutions, the portion of credit portfolio that goes to agricultural activities varies considerably.

Among the member countries of the West African Economic and Monetary Union, a study on rural finance conducted in 200 found that only 14% of overall credit supply goes to the agricultural sector.

Remarkably, 92% of this supply came from the commercial sector, largely surpassing the participation of development banks (5%) and microfinance institutions (3%), whose credit supply is essentially characterized by short-term loans. In India, between 2000 and 2007, only 8% of credits disbursed by the microfinance sector, dominated by the Self-Help Group model, went to agriculture. The rest went to livestock activities (14%) consumer loans, microenterprises and trading activities (78%).

Despite the relatively general and heterogeneous nature of the data available, often only referring to certain geographical areas, there is a consensus that agriculture is inadequately funded and that the supply does not met the needs of farmers.

This is essentially due to the fact that financing agriculture is generally more costly, risky and less profitable than other forms of financing. Beyond the commonly cited difficulties that come with setting up financial services in rural zones, agricultural activities are characterized by a number of specificities that financing mechanisms must take into account.

5.1.3 Specificities of agricultural finance

The agricultural sector is different from other economic sectors in a number of ways. Activities are generally located in isolated areas with low population density and poor infrastructure. They are dependent on weather and production cycles; income is seasonal and monetary income is limited. Agricultural prices are notoriously volatile and few farmers can offer guarantees that are legally or financial acceptable.

These specificities demand financing mechanisms adapted to the diverse needs and services of rural households:

- ❖ Short-term: input financing at the beginning of the crop year (seeds, fertilizers, pesticides), additional labor, feed, storage facilitates, processing, etc.
- ❖ Medium and long term: equipment for intensification, commercialization (transportation), storage (buildings), perennial crops (investment, renewal, maintenance), (re)constitution of herds, land purchase.
- ❖ Family needs: personal, durable goods, housing.
- ❖ Savings
- ❖ Non-financial services: monitoring demand, technical assistance and extension.

Understanding how to best meet these financial needs and finding ways to mitigate the risks associated with them are added challenges that further hinder the expansion of financial services for agriculture. Moreover, as microfinance is increasingly integrating into conventional financial markets, the sector has no choice but to apply cost-covering interest rates. Such rates often contradict the expansion of rural coverage and agricultural finance due to the low profitability of the activities financed. All these factors explain the relative lack of interest in agriculture on the part of urban and peri-urban zones. Consequently, liberalized markets coupled with contractual innovations—elements promoted under the new paradigm—have not fulfilled their promises vis a vis rural and agricultural finance.

5.1.4 A deepening of the new rural finance paradigm

The microfinance sector, motivated in part by the broader trend towards financial inclusion, is changing. New actors are emerging along side the "traditional" ones (NGOs, governments, development banks, peasant organizations, etc.). In addition to commercial banks, one observes increasing involvement of specialized microfinance funds, international guarantee funds, agro-food industries, as well as other agricultural value chain stakeholders.

Given this context and in light of agriculture's specificities, there is a need to take a step forward: identify intermediary channels that encourage interaction between the diversity of public and private actors, old and new. Indeed, while recent changes to the microfinance sector have created a certain "imbalance," they also bring great potential. To take advantage of new opportunities, there is a need for a pluralistic, inclusive approach to the financial sector as a whole and to public/ private partnerships. Conference discussions pinpointed several ways to respond to agriculture's specificities and improve MFIs' provision of agricultural finance:

- Initiatives that bring together diverse stakeholders
- Active partnerships with peasant organizations
- Strong public policies

5.1.5 Testing of New Forms

MFI governance

The idea that one organization model is particularly well-adapted to financing agriculture does not hold up in reality. The institutional type does not appear to be a determining factor of financial access. Indeed, conference discussions identified a variety of models that have proved successful in financing agriculture:

- Self-Help Groups,
- Development banks
- Cooperative (mutualist)-type MFIs or MFIs funding with private capital

For example, development banks are making comeback, where they are once again becoming important players in agricultural finance: 32 of 198 development banks finance agriculture, especially small producers. Among these 32, half allocate 50% of their portfolios to the agricultural sector. In Niger Delta region, three MFIs of varying institutional types and two banks finance 80% of all farmers living in the zone. The cooperative network RCPB dedicates 25% of its loan portfolio to agriculture.

Development banks, for instance, often deal with government interference, which can become a weakness if not well-managed. Banrural S.A. in Guatemala has been successful in this respect, especially given its profitability and penetration rates.

Management has effectively achieved a balance among its different shareholders: government, cooperatives, indigenous and peasant organizations and non-profit associations.

Nonetheless, the level of participation of clients/members in defining products and services in institutional types that ostensibly include them in the governance structure remains limited. In Mexico, farmer representatives play a political role but are not involved in designing financial services. Again, the experience of Banrural S.A. Guatemala is an exception, having achieved genuine convergence between the involvement of clients in strategic decisions and the commitment of managers to meet client demands as a way to ensure the sustainability of services offered.

The conference presentations highlighted the importance of involving clients in the design of financial services to better adapt them to their needs. In this respect, the cooperative models, which bring together salaried staff and elected officials, are conducive for exchanging information and therefore adapting products more effectively. The success of Agricultural Cooperative Bank of Armenia-Crédit Agricole in Armenia demonstrates the crucial role of farmer-members.

Implementing product innovations requires the support of well-trained salaried staff/credit officers who have a thorough understanding of the context in which they work. It is also important that the MFI is genuinely committed to its mission of serving the agricultural sector, since achieving this goal demands the commitment of all employees. However, training or hiring qualified staff also comes at a cost, and raises the question of how to assume these high human resources expenses while pursuing a strategy of financial sustainability.

There are limits, however, to the extent which institutions are able to diversify products and services. Taking once again the example of the Niger Delta in Mali, one observes that the inability to mobilize funds for term loans results in a lack of products to meet the equipment needs of farmers.

Savings offers a solution in this context, and deposit mobilization efforts must be expanded. This calls for MFIs to serve as genuine financial intermediaries, so that deposits can be used not only to guarantee farmers' profits but also to leverage credit portfolios. Rural populations have considerable capacity to save, even if this capacity is not recognized by the banking sector. Microfinance institutions have many advantages that they need to consolidate. The cooperative networks in West Africa are particular active in mobilizing saving compare to the rest of the world.

The current difficulty for MFIs to access medium and long-term credit lines calls for innovative partnerships with private sector players, such as specialized funds. At the same time, while private sector actors can offer solutions, their demands in terms of profitability must be assessed so as not to jeopardize the overall objective of financing agricultural activities.

For private actors to play an important role in agricultural finance and development, there is a need to have a long-term vision. Short-term profitability goals are simply not compatible with financing the agricultural sector.

Generally speaking, the governance of these institutions must bring together proximity, innovation, diversification and risk management. There is a consensus that all the actors in the agricultural sector (financial institutions, producers, buyers, processors) face risks that justify joint efforts to reduce these risks. It is increasingly common to observe the formation of strategic alliances to think collectively about ways to mitigate risk, leverage the comparative advantages of each actor and stimulate knowledge exchange.

5.1.6 Strategic alliances

As discussed, rural financial institutions are using three main strategies:

- diversification of product lines: combining different financial products for agriculture, such as leasing and warrantee;
- diversification of activities financed: including agricultural, non-agricultural and productive activities as well as "social" and consumer expenses;
- formation of strategic alliances: between MFIs, banks and peasant organizations, as well as between MFIs, private sector players and the public sector (government, donors).

These three strategies, which may be pursued in tandem, aim to improve agricultural finance by meeting three important needs: medium- and long-term finance, by increasing credit ceilings for larger equipment loans (thus boosting means of production); long-term partnerships; and the integration of microfinance into rural development dynamics.

Product innovation is necessary but roll-out capacities are still limited

Implementing innovative financial products and, more specifically, term finance involves a number of challenges: client guarantees, trained MFI staff, availability of refinancing (because members' savings are not always enough to finance short- and medium-term loans) and regulatory issues regarding sanctions in case of default (seizure of property or goods, etc.).

Medium- and long-term loans require guarantees that reflect the specific nature of assets held by rural households. Microfinance institutions need to be able to analyze the medium-term repayment capacity of rural households and profitability of agricultural activities. This requires skills in risk and market analysis as well as access to information that is not always easily available.

The issue of profitability is key, because it relates to the securization of agricultural revenue, which involves the stabilization of agricultural prices. Reducing price volatility makes it possible to predict prices, thus removing one of the causes of non-repayment and eliminating one of the risk factors in agricultural finance. Trying out new products and services thus requires MFIs to work collaboratively with a variety of actors and think in terms of synergy and alliances.

The conference revealed that sector players have a genuine interest in meeting these challenges by creating synergies that pool skills and reduce risks. This objective requires a holistic vision of the agricultural sector and involves a wide variety of actors: not only MFIs, peasant organizations, banks and governments but also the private sector (agro-industries, guarantee funds, etc.).

Current partnership dynamics

Overall, most new partnerships revolve more around innovative processes than innovative products. They are emerging at different levels:

a) Financial services: funding, guarantees and dissemination Whether a peasant organization or MFI, guarantees are necessary to access financial markets. The FOGAL (Fonds de Garanties pour l'Amérique Latine) was created precisely to meet this need. Created with support from the Belgian cooperation agency and European Commission, the fund is invested in a European bank and guarantees loans taken by actors.

b) Non-financial services: technical assistance, information, the effectiveness of financial services partly depends on the non-financial services offered to improve agricultural production. One successful way to better support rural and agricultural populations is to create alliances with other actors (NGOs, governmental entities, peasant organizations, etc.) to set up complementary services like training and technical assistance (TA). Training possibilities are endless, ranging from management or financial advisory services to farmers to capacity building for elected representatives of MFIs.

The agricultural package model of DECSI (Debit Credit and Savings Institution) in Ethiopia is an example of an institution offering TA, market information and producer networking services along with its financial services. The approach federates a variety of actors to help boost the development of activities financed.

c) Market access: processing and distribution channels In recent years, it linkages between the microfinance and agro-industrial sectors have become increasingly common, in order to consolidate the comparative advantages of each and encourage exchange.

- MFIs have the expertise, systems and technology required to sustainably provide a range of financial products.
- Agro-industrial firms know the end-consumers, culture, prices, markets and production constraints. Sometimes, they already have input distribution networks in place (involving credit) with farmers.

Several initiatives aiming to strengthen interaction between these two sectors are underway. The objective is to build long-term relationships and reduce risk for the different actors: producers, borrowers, buyers and processors.

These partnerships come in different forms.

- Some focus on one link in the value chain.
- Others address the chain as a whole.
- Partnerships may be direct or indirect, i.e.: incited by a third party, such as an NGO, who plays the role of catalyst, facilitator and sometimes service provider.

There are recent examples of value chain actors playing the limited role of "virtual guarantor", in which case a producer's mere association with a large buyer or processor, for instance, serves as a sign of creditworthiness in the eyes of financial institutions. The value chain actor may also be directly involved in financial transactions, providing producers credit services, a more traditional approach.

5.1.7 Peasant Organisations

The liberalization of agricultural production that has come in the wake of several decades of government intervention has led to a dismantling of state-run agricultural extension, financing and commercialization services. The government's withdrawal has created a gap in the provision of intermediation services and support to farmers, to help them better adapt to the new economic context. Professional agricultural organizations (PAOs) have thus emerged, organizing their activities around two major issues:

- advocacy of agricultural interests and farmer's rights
- provision of support services to their members.

Strategies used by professional agricultural organizations

Access to finance is crucial for professional agricultural organizations (PAOs), constantly confronted with financial constraints. Yet PAOs face major challenges, as financial institutions are reticent to lend to them, or are altogether absent from agricultural regions. To address this situation, PAOs are using a variety of strategies to improve their financial access:

- Internalizing credit operations;
- Creating a "sister" microfinance institution;
- Partnering with existing financial institutions of these three, the partnership strategy appears the most common among PAOs, and is worth exploring further.

Partnerships between PAOs and financial institutions, especially MFIs, offer great potential for agricultural finance. But in practice, relationships between PAOs and MFIs are difficult. Cost of services, for instance, remains a point of contention between the two. To achieve sustainability, MFIs have no choice but to charge cost-covering interest rates. Farmers consider these rates too high, however, given the low profitability margins of their agricultural activities.

Strengthening partnerships between PAOs and MFIs thus requires both parties to overcome a certain number of challenges, not the least of which is determining the type of governance structure to set up, given the diversity of actors involved. Regardless of the approach, the challenge is for MFIs to adopt a governance style that enables all the stakeholders to meet on common ground. This requires an understanding of the differences between each group of stakeholders. PAOs and MFIs work with populations whose expectations and capacities differ. Moreover, within a group of PAOs, there are variations in terms of capacity to negotiate, interact and be involved with an MFI. Similarly, MFIs do not constitute a homogeneous group, either.

It is also important that the governing stakeholders in these partnerships understand the risk of mission drift and find ways to ensure the MFI's long-term commitment to serving small producers. In this respect, a well-designed governance structure can help avoid abuse or infringement of the different stakeholders' rights.

This strategy effectively "outsources" the financial service, thus raising the issue of how to sustain the non-financial services offered by the PAO. Conference discussions emphasized the importance of these services in the development of agricultural activities. A mix of non-financial production factors is required for finance to be successful.

Operational guide to strengthen PAO/MFI partnerships

Generally speaking, capacity building of the various actors—elected officials, managers and extension officers—is a critical success factor for PAO/MFI partnerships. This observation incited the development of an operational guide to help strengthen such partnerships). The guide is being finalized by members of the CERISE network, with the financial support of FARM Foundation.

There appears three hypotheses:

- ❖ Partnerships between PAOs and financial institutions (FIs) can improve outreach of agricultural finance;
- ❖ Due to the distance between these two types of entities, partnerships are not spontaneous and must be actively built;
- ❖ Consolidating the knowledge and skills of the two institutions can help foster partnerships.

The specific objectives are:

- To improve PAOs' and FIs' capacities to analyze the financial needs and constraints of POs and their producers;
- To help PAOs choose efficient and viable ways to access financial services;
- To give PAOs and FIs the tools to build partnerships and improve POs' access to financial services.

Although building partnerships between PAOs and MFIs represents progress, a number of issues remained unresolved: finding ways to mobilize resources adapted to agricultural finance, especially equipment needs; managing agricultural risk; developing financial services in particularly isolated zones and addressing what are perceived to be relatively high interest rates, compared to profitability levels.

None of these issues can be effectively or equitably resolved by the market or negotiations between economic actors. They call for well-adapted redistribution policies and incentives. FONGS's decision to become a shareholder in the Caisse Nationale de Crédit Agricole du Sénégal (putting up 4% of capital) was guided by the desire to have a seat on the Board of Directors so as to secure the only agricultural finance mechanism available, and influence lending policies. This strategy has led to a number of victories, including:

- Expansion of CNCAS's network and increased proximity to rural producers;
- Reduction of interest rates from 18% to 7.5%;
- The beginning of a dialogue regarding linkages between CNCAS's network and decentralized endogenous savings and credit cooperatives. It is an example of how the joint collaboration of actors at various levels can help ensure financial access for farmers.

5.1.8 Renewing Public Policy Incentives

The withdrawal of government interventions in developing countries has not resulted in expanded banking services in isolated areas, nor has microfinance been able to fill the vacuum. Supply is still insufficient. In rural and agricultural regions where favorable conditions render some agricultural activities profitable, MFIs are more likely to achieve sustainability, and are thus open to financing agriculture. Inversely, MFI penetration is much more limited (or altogether absent) in marginalized regions where running a cost-covering institution is more difficult.

There is a growing consensus that market mechanisms are not enough to encourage a well-adapted supply of rural, and especially agricultural, finance and the debate on the role of public authorities in improving financial access for farmers is once again open.

5.2 KCC

5.2.1 Introduction

Agricultural finance can be dealt with both micro level and macro level. Macro-finance deals with diverse resource of arising finances for agriculture as a entire in the economy. It is also concerned with the loan process, rules, system, monitoring and controlling of diverse agricultural credit organizations. Hence macro-finance is associated to financing of agriculture at aggregate level.

The Kisan Credit Card (KCC) scheme is a credit scheme introduced in August 1998 by Indian banks. This model scheme was prepared by the National Bank for Agriculture and Rural Development (NABARD) on the recommendations of R.V.GUPTA committee to provide term loans and agricultural needs.

Its objective is to meet the comprehensive credit requirements of the agriculture sector by giving financial support to farmers. Participating institutions include all commercial banks, Regional Rural Banks, and state co-operative banks. The scheme has short term credit limits for crops, and term loans. KCC credit holders are covered under personal accident insurance up to 50,000 for death and permanent disability, and up to 25,000 for other risk. The premium is borne by both the bank and borrower in a 2:1 ratio. The validity period is five years, with an option to extend for up to three more years. Kisan Credit Card (KCC) offering credit to the farmers in two types viz,

1. Cash Credit

2. Term Credit (for allied activities such as pump sets, land development, plantation, drip irrigations).

Due to lack of awareness among farmers and unnecessary delays, cumbersome procedure and improper practices adopted by institutional lending agencies; a large number of Farmers heavily depend on non-institutional sources of credit for their frequent needs to purchase farm inputs such as seeds, fertilizers, pesticides etc. The non-institutional credit is not only expensive but also counter-productive. The Kisan Credit Card scheme was launched to provide adequate, timely and cost effective institutional credit from the banking system to the farmers for their cultivation needs. Farmers can not only purchase inputs but also can withdraw cash from this credit card for their input needs.

5.2.2 Progress

Kisan Credit Cards are issued to the farmers on the basis of their land holdings and other criteria such as timely payment of past credits etc. Farmers covered under the Kisan Credit Card scheme are issued with a credit card and a pass book or a credit card cum pass book incorporating the name, address, particulars of land holding, borrowing limit, validity period, a passport size photograph of holder etc., which may serve both as an identity card and facilitate recording of transactions on an ongoing basis.

The innovation of Kisan credit card scheme is proved to be very successful. As per NABARD, as on March 31, 2012, the banking system has issued 11.34 crores KCCs and sanctioned loans up to Rs. 5, 21, 617 crores since the inception of the scheme in 1998-99.

As on 31 March 2010, maximum number of KCCs issued in Uttar Pradesh followed by Andhra Pradesh. Thus, these two States together accounted for one third of the total KCCs issued so far. Similarly, at end-March 2010, Uttar Pradesh had the maximum share of loans sanctioned under the KCC scheme followed by Maharashtra. However, as on 31 March 2010, average amount of loan sanctioned per KCC was the highest in Gujarat followed by Punjab.

As on 31 March 2011, among the major banking institutions, commercial banks, cooperative banks and regional rural banks accounted for about 48.59 per cent and 35.23 per cent and 16.17 per cent of the total number of cards issued respectively. In terms of total loan disbursed to cardholders, the share of commercial banks was 63.31 per cent, followed by 17.62 per cent for cooperative banks and 12.77 per cent for RRBs. Coverage of marginal farmers and small farmers in the KCCs were in the range of 63-68 per cent (Cooperative banks), 58-61 per cent (RRBs) and 59-64 per cent (CBs). Share of tenant farmers was less than one per cent.

In Eastern Rajasthan, over 92 per cent of agricultural credit for working capital needs is met by borrowing from family, friends, money lender and primarily through own funds. Only eight per cent is from banks. In Southern Rajasthan, over 97 per cent of agricultural credit for working capital needs is met by borrowing from family, friends, money lender and primarily through own funds and only two per cent is from banks. In central Rajasthan, over 93 per cent of agricultural credit for working capital needs is met by borrowing from family, friends, money lender and primarily through own funds and only seven per cent is from banks. In Northern Rajasthan and Western Rajasthan conditions is very worst as compared to other regions.

In Karauli district of Rajasthan total number of villages is 881 and six tehsils. The population of Karauli district is 14, 58, 459 persons and literacy rate is 67.34 per cent. Total geographical area is 5, 04, 301 ha in which net sown area is 1, 52, 734 ha, forest area 1, 72, 632 ha, permanent pasture 30, 974 ha and Non-agriculture 85, 172 ha. Total number of banks is 63 in which 39 Commercial banks, 13 RRBs, 6 Cooperative banks and 5 Land development banks.

5.2.3 Features

The Kisan Credit Card is a credit card to provide affordable credit for farmers in India. It was started by the Government of India, Reserve Bank of India (RBI), and National Bank for Agricultural and Rural Development (NABARD) in 1998-99 to help farmer's access timely and adequate credit. The Kisan Credit Card allows farmers to have cash credit facilities without going through time consuming bank credit screening processes repeatedly. Repayment can be rescheduled if there is a bad crop season, and extensions are offered for up to four years. The card is valid for three years and subject to annual renewals. Withdrawals are made using slips, cards, and a passbook.

Credit to farmers or groups of farmers for crop loans, working capital or investment credit for viable agriculture purpose. All farmers, owner cultivators, tenant cultivators, share croppers and individual farmers having agreement with institution are eligible for the card. Most banks in India that offer agricultural finance offer the Kisan Credit Card. Under the scheme, Branches may issue Kisan Credit Cards to the farmers who are otherwise eligible for sanction of short term credit for crop production, allied activities and other nonfarm activities. The farmers should come from the operational area of the Branch. Credit is given as per scale of finance specific to the crop and KCC norms.

The interest rate is 7% (SI), out of which 3% subsidy is given by Government of India if aggregate credits are more than aggregate debits in a year. The card is valid for five years, of which crop loan and working capital components have to be renewed annually. Maximum limits: Rs. 50000/- for Rabi Crops and Rs. 50000/- for Kharif Crops.

Since, the scheme of KCC is under implementation by State Cooperative Banks (SCBs) through DCCBs and PACS as also the Regional Rural Banks (RRBs) and Commercial Banks (CBs) under the guidelines of NABARD. As on 31 March1999 total number of KCC issued was 7.84 lakh by various banks. In the initial years of incorporation of KCC Scheme Co-operative banks have the largest share (62%), followed by commercial banks (30%) and RRBs (8%). The performance in the implementation of the KCC scheme has been impressive in the States of Andhra Pradesh, Gujarat, Haryana, Karnataka, Maharashtra, Punjab, Rajasthan, Tamil Nadu, Uttar Pradesh and Uttaranchal.

The Kisan Credit Card Scheme was a step towards facilitating the access to shortterm credit for the borrowers from the formal financial institutions. The scheme was conceived as a uniform credit delivery mechanism, which aimed at provision of adequate and timely supply of short-term credit to the farmers to meet their crop production requirements. The KCC instrument would allow farmers to purchase agriculture inputs such as seeds, fertilizers, pesticides and also allow them to withdraw some cash for meeting their other crop production related requirements.

Features of Kisan Credit Card Scheme

- Farmers eligible for production credit of Rs. 5000 or more are eligible for issue of Kisan Credit Card.
- Eligible farmers to be provided with a Kisan Credit Card and a pass book or card-cum-pass book.
- Revolving cash credit facility involving any number of drawls and repayments within the limit.
- Limit to be fixed on the basis of operational land holding, cropping pattern and scale of finance.
- Entire production credit needs for full year plus ancillary activities related to crop production to be considered while fixing limit.
- Sub-limits may be fixed at the discretion of banks.
- Card valid for 3 years subject to annual review. As incentive for good performance, credit limits could be enhanced to take care of increase in costs, change in cropping pattern, etc.
- Each drawals to be repaid within a maximum period of 12 months.
- Conversion/re-scheduling of loans also permissible in case of damage to crops due to natural calamities.
- Security, margin, rate of interest, etc. as per RBI norms.
- Operations may be through issuing branch (and also PACS in the case of Cooperative Banks) through other designated branches at the discretion of bank.
- Withdrawals through slips/cheques accompanied by card and passbook.

5.2.4 Benefits of Kisan Credit Card Scheme

- Simplifies disbursement procedures
- Removes rigidity regarding cash and kind
- No need to apply for a loan for every crop
- Assured availability of credit at any time enabling reduced interest burden for the farmer.
- Helps buy seeds, fertilizers at farmer's convenience and choice
- Helps buy on cash-avail discount from dealers
- Credit facility for 3 years – no need for seasonal appraisal
- Maximum credit limit based on agriculture income
- Any number of withdrawals subject to credit limit
- Repayment only after harvest
- Rate of interest as applicable to agriculture advance

- Security, margin and documentation norms as applicable to agricultural advance
- Access to adequate and timely credit to farmers
- Full year's credit requirement of the borrower taken care of.
- Minimum paper work and simplification of documentation for drawal of funds from the bank.
- Flexibility to draw cash and buy inputs.
- Assured availability of credit at any time enabling reduced interest burden for the farmer.
- Flexibility of drawals from a branch other than the issuing branch at the discretion of the bank.

5.2.5 Growth and relevance of KCC and Micro-financing in India

As per financial innovation in Indian agricultural credit market: progress and performance of kisan credit card observed that kisan credit card (KKC) as a financial product is becoming very popular and successful. The study has identified the socio-economic factors that affect the farmer's access to KCC scheme. The effect of education has indicated the need for capacity building of the farmers. Imparting training to borrowers regarding procedural formalities of financial institutions could be helpful in increasing their access to KCC scheme. However, the bias and prejudice on the basis of land-holding size and social castes should be addressed on a priority basis. The purpose behind launching of this innovative product is to facilitate adequate and hassle-free credit to all the farmers in times of need. Therefore, concerted efforts will have to be made at different levels to further popularize the KCC scheme.

As per the study of issuance of kisan credit card scheme, it was reported that over 14 million KCCs have been issued across India. The scheme has been implemented in all the states and union territories by 27 public sector banks (PSBs), 373 district central co-operative banks (DCCBs) and 196 Regional Rural Banks (RRBs). The KCC system played a vital role in rural credit and large numbers of farmers have been benefited by this novel and innovative credit system. It was the duty of farmers to utilize this credit delivery system not only for their individual benefit but also for national development at large.

As see, the kisan credit card serves as a boom for small farmers through use of KCC by co-operatives, RRBs and commercial banks in different years which results in high demand and necessity. Kisan credit card scheme was actively functioning in the initial years, after that there was some set back. The Central Government introduced KCCs to provide loan facilities to small and marginal farmers in India. The study revealed that under the scheme more than three crores of farmers were benefited and the scheme could be extended to other farmers in the coming years.

Based on the idea of supply analysis of institutional credit to agriculture for various states across India, it was analyzed that inter-state disparities exists during the flow of agricultural credit. The study reveals that the growth of agricultural credit was higher during pre-reform period compared to the reform period in most of the states. It is also observed that the growth rate of agricultural credit was uneven during the sub-periods as well as across the states. Since the ABCD has been declining for all the major states, it suggests that bankers are utilizing lower and lower amounts of deposits mobilized for agricultural lending. The analysis on access to formal credit and the outstanding loan amount per account reveals that there are inter-state differences in the access to credit as well as the loan amount obtained by the farm households even for the same size class of landholding.

The performance of rural credit and factors affecting choice of credit sources assessed the performance of credit flow which identified the factors that influenced the choice of credit outlet and the possession of kisan credit cards by rural households. Using NSS data for the years 1991-92 and 2002-03, it was revealed that access and distribution of rural credit, in general, increased but was skewed in favour of more developed states and within these towards better-off households. The poorer households were more dependent on non- institutional sources of finance. The use of kisan credit card was encouraging and its distribution was less skewed. Age, male-headed households, household size, farm size, level of education and self-employment in agriculture appeared as significant variables positively determining the choice of institutional sources of credit and possession of kisan credit cards.

In India, the adaptation of KCC in Uttar Pradesh and among different regions and social classes was not very good. It was observed that since there was uneven development of different regions and social classes, the access to one of the most important innovation (KCC) was not even throughout the state. Only land and household size had positive relationship with probability of a person of being a KCC holder. He suggested that, if the waste land was made cultivable to weaker section of society then there might be considerable improvement with respect to even distribution of KCC in the state.

As per the study on kisan credit card: missing target reviewed the status of KCC in Madhya Pradesh and concluded that only 40 per cent of total 65 lakh farmers have been distributed KCCs in the state against target of covering all the farmers by March 31, 2004. Except bank of Rajasthan no private bank had issued any KCC to the farmers. Out of 4, 17,314 cards issued, commercial banks issued 1, 80,745 cards, RRBs issued 63,002 cards and co-operative banks issued 1, 73,567 cards.

The cash benefits of kisan credit card scheme are upon farmers to analyzed the functioning of KCC in various areas of Punjab state. During the study they found that at all India level the cumulative number of KCC issued by all implementing agencies as on March 31, 2004 stood at 413.19 lakhs and corresponding figure for Punjab was

13.78 lakhs. The Co-operative banks accounted for 65 per cent of KCC issued and 63 per cent of amount sanctioned. The commercial banks (PSBs and RRBs) accounted for 35 per cent of KCC issued and 37 per cent of loan sanctioned. The Punjab topped the list with highest penetration ratio.

As per the examination of status of KCC in India and Kerala, it was reported that a total of 28, 16,437 KCCs amounting to Rs. 3,108 corers were issued by 293 banks of which 152 were Central Co-operative Banks (51.88 %) and 141 were RRBs (48.12 %). The Central Co-operative Banks issued 27, 69,000 KCC against which Rs.2, 973 corers were sanctioned (95.66 %). The RRBs issued 47,437 cards amounting Rs. 135 corers (4.34%). The study also revealed that 86 per cent of KCC holders withdrew the amount as ready cash and remaining by means of chaques and utilised it mainly for agriculture purpose particularly for purchasing fertilizers and for paying labour charges.

Patel (1999) elaborated features and need for KCC in India and recommended that bankers committees both at the block level and state level should monitor not only the number of KCC holders on quarterly basis but also examine the amount sanctioned, disbursed and utilised which have direct impact on crop production in the country. Besides feedback, utility and effectiveness of KCC scheme might be looked into in order to modify it to fit the needs of users than introducing uniform scheme in the country.

The case study based on borrowers knowledge on farm credit and follow up action of bank officials in Khajipet mandalam in Cuddaph district of Andhra Pradesh shows that 54 per cent of borrowers opined that scale of finance provided by bank was sufficient but all the beneficiaries wanted technical guidance. 60 per cent of the borrowers were of the opinion that loan sanctioning procedure was easy and 82 per cent of the borrowers were of the opinion that interest charged was reasonable. The study revealed that bank personnel have visited short term loan beneficiaries once during the crop season. It was suggested that timely advance should be provided and bank authorities should change the procedure to suit local conditions and also for providing technical guidance to borrowers.

As examined, the credit position for the poor in India, numerous programmes have been designed specifically to enlarge the flow of credit to the poor through programmes like IRDP, differential rate of interest, self employment programmes. Despite these efforts, the desired effect had not been achieved. The cumbersome lending procedures, inadequate supervision and the apathy of bank staff in handling rural credit resulted in delayed and untimely credit, which further led to large scale mis-utilisation of loans and default of credit.

5.3 Lead bank scheme

5.3.1 Introduction

The Lead Bank Scheme, introduced towards the end of 1969, envisages assignment of lead roles to individual banks (both in public sector and private sector) for the districts allotted to them. A bank having a relatively large network of branches in the rural areas of a given district and endowed with adequate financial and manpower resources has generally been entrusted with the lead responsibility for that district. Accordingly, all the districts in the country have been allotted to various banks. The lead bank acts as a leader for coordinating the efforts of all credit institutions in the allotted districts to increase the flow of credit to agriculture, small-scale industries and other economic activities included in the priority sector in the rural and semi-urban areas, with the district being the basic unit in terms of geographical area.

5.3.2 Features

The Lead Bank Scheme (LBS) was introduced by Reserve Bank in 1969 when designated banks were made key instruments for local development and entrusted with the responsibility of identifying growth centres, assessing deposit potential and credit gaps and evolving a coordinated approach for credit deployment in each district, in concert with other banks and other agencies.

The LBS underwent significant transformation in 1989 when the Service Area Approach was dovetailed into the scheme. Subsequently, as it was observed that the service area restrictions were a limiting factor for credit deployment, the restrictive provisions were removed in 2004, except for the Government Sponsored programmes. As at March 2009, there were 26 banks, mostly in the public sector, which have been assigned lead responsibility in 622 districts of the country.

5.3.3 Origin and Objectives

The genesis of Lead Bank Scheme can be traced to the Study Group, headed by Prof. D. R. Gadgil (Gadgil Study Group) on the Organisational Framework for the Implementation of the Social Objectives, which submitted its report in October 1969. The Study Group drew attention to the fact that commercial banks did not have adequate presence in rural areas and also lacked the required rural orientation.

As a result, the banking needs of the rural areas in general and the backward regions in particular, could net be adequately taken care of by the commercial banks and the credit needs of rural sector of the economy, particularly agriculture, small-scale industry and services sectors remained virtually neglected.

The Study Group, therefore, recommended the adoption of an 'Area Approach' to evolve plans and programmes for the dev cover the metropolitan cities of Mumbai, Delhi, Kolkata, Chennai and certain Union Territories.

Objectives

1.Eradication of unemployment and under employment

2.Appreciable rise in the standard of living for the poorest of the poor

3.Provision of some of the basic needs of the people who belong to poor sections of the society

5.3.4 Service Area

By late eighties, studies undertaken by RBI to assess the impact of bank credit in increasing production, productivity and income levels of the rural population revealed certain weaknesses in the system of dispensation of rural credit, viz. the rural lending of bank branches was haphazard and dispersed in a large number of villages spread over a wide area rendering supervision difficult.

Absence of effective local level planning taking into account the potential for development, poor availability of infrastructure and linkages etc. were identified as the contributory factors. It was, therefore, felt necessary to adopt an approach of assigning specific service areas to the bank branches in rural and semi-urban areas, paving the way for 'Service Area Approach' to rural lending.

Service Area Approach (SAA) basically aimed at planned and orderly development of an identified command area which would enable the branch to have development orientation and concentrate on productive lending, thereby contributing to the development of specific areas assigned to it. The Union Finance Minister in his budget speech for the year 1988-89 had also referred to the proposed f dispensation under which each bank branch would have a designated service area.

In order to examine the operational aspects involved in the implementation of this approach, a Committee (Chairman: Dr. P. D. Ojha) was set up by the Reserve Bank. On the basis of the recommendation of the Committee, a decentralized planning policy was adopted and Service Area Approach to Rural Lending was introduced with effect from April 1, 1989, involving five distinct stages in its implementation:

- Identification of the service area for each bank branch;
- Survey of the villages in the service area for assessing the potential for lending for different activities and identification of beneficiaries for assistance;
- Preparation of credit plans on an annual basis for the service area by each branch;
- Co-ordination between credit institutions on the one hand and field level development agencies on the other on an on-going basis for the effective implementation of credit plans;

- ❖ A continuous system of monitoring the progress in the implementation of the plans and individual schemes.

The above approach to rural lending was intended to bring about a major change in the quality and productivity of rural lending and forge effective linkages between bank credit, production, productivity and increase in income levels.

The basic principle of SAA was demarcation of service area and preparation of credit plans for systematic administration of credit in the assigned area. Under the SAA, all rural and semi-urban branches of commercial banks and regional rural banks were allocated specific number of villages (about 15 to 25 villages), generally in geographically contiguous areas and proximity to the branch concerned, the credit needs of which were to be taken care of by the respective service area branches.

It was also to be ensured that the designated area of a bank branch was not intercepted by the designated area of another bank branch. The service areas were allocated to the branches by a Committee headed by the Lead District Officer of the Reserve Bank of India with Lead Bank Officer of the district and a representative of NABARD as members.

In 1989, the Service Area Approach (SAA) was adopted wherein villages were identified and assigned to bank branches based on their proximity and contiguity and by adopting a cluster approach. Credit plans were prepared on an annual basis for the service area of each branch which involved co-ordination between the various developmental agencies and credit institutions. Due to allotment of villages to designated bank branches, the activities of the 'service area branches' were restricted to the allotted villages and they were unable to provide financial assistance outside their service areas, despite being in a position to do so.

Similarly, borrowers belonging to these villages were required to approach the 'designated bank branches' for their credit needs and were not in a position to avail of services of any other bank branches, irrespective of whether they were satisfied with the services provided by the designated bank branches or not. The Advisory Committee on Flow of Credit to Agriculture and Allied Activities observed that the Service Area Approach, introduced for planned and orderly development of rural areas, had developed rigidities and acted as a bottleneck despite built in measures to provide flexibility.

The Committee recommended that this feature of flexibility needed to be preserved and the service area concept made mandatory only for government sponsored schemes. Pursuant to the recommendations of the Committee, the restrictive provisions along with certain other aspects of service area were removed from December 2004, except for the government sponsored schemes.

5.3.5 Organization Structure

The organization structure of Lead Bank Scheme can be broadly divided into following levels.

1. State Level Committees
2. District Level Committees
3. Block Level Committees

1. State Level Committees

The state level committes under Lead Bank Scheme are the APEX bodies that control the execution of the scheme. State level committees consists of

- State Level Bankers Committee (SLBC)
- State Level Coordination Committee (SLCC)
- Steering Subcommitties (SC)

State Level Bankers' Committee (SLBC)

This is one of the most important fora under the Lead Bank Scheme and has proved highly effective in quickly implementing policy decisions especially under emergent situations like natural calamities. The State Level Bankers' Committees (SLBCs) have been constituted, following instructions from Government of India in April 1977, as an inter-institutional forum for co-ordination and joint implementation of development programmes by all financial institutions operating in a State.

It comprises the representatives of commercial banks (including RRBs), State Cooperative Banks (StCB), Land Development Banks (LDB), NABARD, RBI, etc. Representatives of various organizations from different sectors of the economy like retail traders, exporters, agricultural graduates (who wish to start agri-clinics/ agribusiness), farmers' unions and some other organizations, who wish to protect the interest of their members, may also be invited as special invitees in SLBC meetings for discussing their specific problems.

The meetings are chaired by the Chairman/Executive Director of the convenor bank. The forum reviews the banking developments in the State with special reference to Annual Credit Plans, Government sponsored programmes, flow of credit to priority sector, branch expansion, Credit Deposit ratio, etc.

The SLBCs are also expected to take up for consideration, such issues as have been raised by the member banks or by the State Government authorities and questions or inter bank differences of views and approaches remaining unresolved at the District Consultative Committees (DCCs).

The meetings of the SLBC are to be held on a quarterly basis.

- Regional imbalances in availability of banking facilities
- Regional imbalances in availability of banking facilities
- Regional imbalances in deployment of credit
- Liaison with State Government
- Review of functioning of DCC
- Review of implementation of District Credit Plans
- Uniformity in terms and conditions of lending
- Review of credit flow to small borrowers in the neglected sectors, pending loan applications and assistance under various Govt. sponsored programmes
- Review of Credit Deposit Ratio, and
- Any other State specific issues, with the permission of the chair information thereon. In view of these developments, there is a need to make this State level forum compact and more effective.

State Level Co-ordination Committee (SLCC)

The SLCCs which started functioning from mid-seventies were intended to serve as the main forum for co-ordination between the State Government and financial institutions and for the review of the performance of banks particularly in assisting credit based development programmes taken up by the State Government. Meetings of the SLCC are convened by the Chief Secretary, Director of Institutional Finance, Registrar of Co-operative Societies, etc.

They are generally presided over by the Chief Minister/Finance Minister of the State.

The members of SLCC consist of representatives from banks, RBI, NABARD, StCBs, SLDBs and the concerned development departments of the Government.

The functions of the SLCC are to nake in-depth review of banks' performance in the implementation of various credit sased development programmes of Government, matters relating to infrastructure and extension facilities, recovery performance including legislative and administrative support from Government, problems of banking development in the State, etc.

The periodicity of the SLCC is yearly, though, the Standing Committee of 5LCC is expected to meet at quarterly intervals. It has however, been observed that SLCCs are defunct since long in several States and no meetings are being conducted.

Steering Sub-Committee of SLBC

These are functional sub-committees of SLBC to look into specific areas like agriculture, micro, small/medium industries/enterprises, handloom finance, etc.

and evolve a course of action for adoption by the full committee and consist of representatives from few major commercial banks/RRBs, RBI, NABARD, StCB, SLDB etc. The steering sub-committee is expected to meet more frequently than the SLBC.

2. District Level Committees

The District Level Committes under Lead Bank Scheme are the monitoring bodies that control the block level execution of the scheme. District level committees consists of

- District Consultative Committee (DCC)
- District Level Review Committee (DLRC)
- Standing Committee of DCC

District Consultative Committee (DCC)

The District Consultative Committees (DCCs), constituted in the early seventies, marked a significant step towards co-ordination of activities of all commercial banks and other financing agencies on the one hand and Government departments on the other.

This is conceived to be a common forum for bankers as well as government agencies/departments to come to a common platform to find solutions to the problems arising hindering the smooth functioning of the various developmental activities under the Scheme at the district level. All the commercial banks, co-operative banks including DCCB and SLDB, RRBs, NABARD, etc. and various State Government departments and allied agencies constitute the members of the DCC.

The Lead District Manager (LDM) of the lead bank is the convenor of DCC. The Lead District Officer (LDO) of Reserve Bank is also a member of the DCC. The District Collector functions as the Chairman of this committee. Its membership is generally restricted to 20-25 for effective and meaningful discussions in the meeting.

A small functional sub-committee of the DCC is constituted in the districts for monthly monitoring of the government sponsored schemes. Various important matters concerning DCC which required urgent attention are to be discussed in this forum.

District Level Review Committee (DLRC)

Apart from convening DCC meetings, one meeting of DCC every year was held as a District Level Review Meeting (DLRM) with a view to evaluating the progress made in the implementation of schemes included in DCP/ACP, identifying problem areas and devising suitable remedial steps.

The Working Group to Review the Working of the Lead Bank Scheme, observed that the DLRMs had not only served as useful forum for bringing together various organisations participating in the lead bank scheme and non-official agencies connected with rural development but also evolved, in a way, into forums for a general review of performance under DCP/AAP so as to ensure regular participation of all concerned agencies including non-officials in the developmental programmes under implementation in the district.

The Group recommended that the frequency of these meetings may be increased and be held on a half-yearly basis instead of yearly basis, as also designate this forum as District Level Review Committee (DLRC) to reflect the nature of its functioning.

Standing Committee (SC) of the DCC

The Reserve Bank had issued guidelines to banks for formulation of District Credit Plans for 1980-82, on a uniform pattern, which envisaged constitution of a district level task force under the chairmanship of District Collector, comprising representatives from DCCBs, commercial banks having large number of branches in the district and the district planning officials, to assist the credit planning team/ Lead Bank Officer. Subsequently, this Task force was converted into a 'Standing Committee' of District Consultative Committee (DCC) so that the same body which was associated with the formulation of the credit plan could also continuously monitor the progress in its implementation.

The membership of the Standing Committee has been enlarged to include the representatives from RBI, Agriculture Refinance and Development Corporation (ARDC) (now NABARD), DRDA, cooperative department, LDB, etc. RBI has not stipulated any guidelines on the composition of the Standing Committee except that it should be a compact forum which could meet as frequently as might be necessary.

The periodicity of the Standing Committee meetings has also been left to the exigencies of work but it has been recommended that the meetings should be held at least twice in a quarter. With the extension of the IRDP to all the blocks in the country, review of implementation of the Programme had to be undertaken at frequent intervals and accordingly, RBI issued instructions in August 1983 for convening the meetings of the Standing Committees once in a month.

The functions of the Standing Committee, inter alia, covered the following:

- Regular monitoring of Government Sponsored Schemes, and
- Any other important matter concerning DCC requiring urgent attention.

3. Block Level Committees

The Block Level Committes under Lead Bank Scheme are the executive bodies at the grassroot level that control the block level execution of the scheme. Block level committees consists of only one committee, i.e Block Level Bankers' Committee (BLBC)

Block Level Bankers' Committee (BLBC) on implementation of Service Area Approach to Rural Lending, Block Level Bankers' Committees have been constituted in'each block, with the Lead Bank Officer as the Chairman.

This forum aims at achieving coordination between credit institutions on one hand and field level development agencies on the other and helps in the effective implementation of credit plans at the block level. All the banks operating in the block including the district central co-operative banks and RRB, Block Development Officer, and other technical officers in the block, such as extension officers for agriculture, industries and co-operatives are members of the Committee. The Lead District Officer of RBI and the officer concerned from NABARD selectively attend the meetings of the BLBCs. The representatives of Panchayat

Samitis are also invited to attend the meetings at half yearly intervals so as to share their knowledge and experience on rural development in the credit planning exercise.

The main functions of the Committee are as follows:

- To discuss the action plans of different bank branches and their aggregation into Block Credit Plan;
- To resolve operational problems in implementation of the credit programmes of banks, particularly in regard to ensuring availability of inputs and linkages;
- To review the progress in implementation of Government-sponsored schemes;
- To review the implementation of the Block Credit Plan and the provision of other inputs required so that bank credit becomes more productive;
- Allocation of service area of a new branch to be opened in the block as and when licence for a new branches are issued by the Reserve Bank of India;
- Hold meetings at least once in a quarter.

5.3.6 Reasons for Scheme to became inactive

- The Lead bank Scheme was not fully able to achieve its targets due to shift in policies, complexities in operations and issues shifting to the Financial Inclusion.
- Lack of coordination between district planning authorities and banking institutions operating in a district on one side and between Nabard and the Lead Bank on the other is the prominent reason which required attention. Duplication of efforts in credit plan preparation should be avoided by empowering the plan team at the district level appropriately.
- Over the period the system of lead bank scheme and associated district-level coordination committees of bankers has apparently become inactive.

- There was a strong need felt to revitalize the scheme with clear guidelines on respecting the bankers' commercial judgments' even as they fulfill their sectoral targets.
- Various committees like Block Level Bankers Committee, District Coordination Committee and District Review Committee seldom function with all seriousness.
- LBS Information System does not have any checks and balances and does not agree with several other returns relating to Priority Sector Credit.

5.4 RRBs

5.4.1 Introduction

Priority Sector lending includes lending to those sectors that impact large sections of the population, the weaker sections and the sectors which are employment-intensive such as agriculture, and tiny and small enterprises. With a view to providing more credits to the segments under priority sector, RRBs are required to achieve a target of 60% of their outstanding advances for priority sector advances of which at least 25% (i.e. 15% of the total advances) are to be made to weaker sections of the society.

Regional Rural Banks (RRBs) have undergone significant structural and operational changes, be it two-phased amalgamation, implementation of CBS platform or recapitalization, inter alia. Considering the growing significance of RRBs in pursuit of financial inclusion agenda, it has been decided to revise the priority sector guidelines for RRBs.

5.4.2 Categories of Priority Sector

1. a) Agriculture (Direct Finance)

Direct Finance to agriculture includes short term and long term loans given for agriculture and allied activities (dairy, fishery, piggery, poultry, bee-keeping, etc.) directly to individual farmers, Self Help Groups (SHGs) or Joint Liability Groups (JLGs) of individual farmers and to others (such as corporate, partnership firms and institutions), for taking up agriculture/allied activities.

Direct finance includes:-

- Short-term loans for raising crops, i.e. for crop loans including traditional/ non-traditional plantations and horticulture.
- Advances up to Rs. 10 Lakh against pledge/hypothecation of agricultural produce (including warehouse receipts) for a period not exceeding 12 months, irrespective of whether the farmers were given crop loans for raising the produce or not.

- Working capital and term loans including credit sanctioned under Kisan Credit Card for financing production and investment requirements for agriculture and allied activities.
- Loans to small and marginal farmers for purchase of land for agricultural purposes.
- Loans to distressed farmers indebted to non-institutional lenders, against appropriate collateral or group security.
- Loans granted for pre-harvest and post-harvest activities such as spraying, weeding, grading, sorting, processing and transporting undertaken by individuals, SHGs cooperatives in rural areas.
- Loans granted for agricultural and allied activities, irrespective of whether the borrowing entity is engaged in export or otherwise.

Finance to others (corporate, partnership firms and institutions) includes:-

- Loans granted
  - Credit for purchase and distribution of fertilizers, pesticides, seeds, etc.
  - Loans up to Rs. 40 lakh granted for purchase and distribution of inputs for the allied activities such as cattle feed, poultry feed, etc.
- Finance for setting up of Agri-clinics and Agribusiness Centres.
- Finance for hire-purchase schemes for distribution of agricultural machinery and implements.
- Loans to farmers through Primary Agricultural Credit Societies (PACS), Farmers' Service Societies (FSS) and Large-sized Adivasi Multi Purpose Societies (LAMPS).
- Loans to cooperative societies of farmers for disposing of the produce of members.
- Financing the farmers indirectly through the co-operative system (otherwise than by subscription to bonds and debenture issues).
- Loans for construction and running of storage facilities (warehouse, market yards, godowns, and silos), including cold storage units designed to store agriculture produce/products, irrespective of their location.
- Finance extended to dealers in drip irrigation/sprinkler irrigation system/ agricultural machinery, irrespective of their location, subject to:-
  - The dealer should be dealing exclusively in such items or if dealing in other products with distinct records in respect of other items;
  - A ceiling of upto Rs. 30 Lakh to be observed.
- Loans to Arthias (commission agents in rural/semi-urban areas functioning in markets/mandies) for extending credit to farmers, for supply of inputs as also for buying the output from the individual farmers/ SHGs/ JLGs.

- ❖ Credit outstanding under loans for general purposes under General Credit Cards (GCC).
- ❖ Loans granted to NGOs/MFIs for on-lending to individual farmers or their SHGs/JLGs.
- ❖ Overdrafts, up to Rs.25, 000 (per account), granted against 'no-frills' accounts in rural and semi-urban areas.
- ❖ Loans eligible for classification as direct/indirect finance to agriculture
- ❖ Credit under the Kisan Credit Card would be treated as direct finance for agriculture.

Loans not eligible for Classification as Direct / Indirect finance to Agriculture

Loans sanctioned to NBFCs for on-lending to individuals or other entities against gold jewellery, investments made by banks in securitised assets originated by NBFCs, where the underlying assets are loans against gold jewellery, and purchase / assignment of gold loan portfolio from NBFCs.

2. Micro & Small Enterprises

Direct finance to MSE includes:-

- ❖ Manufacturing Enterprises

(a) Micro (manufacturing) Enterprises

Enterprises engaged in the manufacture/production, processing or preservation of goods and whose original cost of investment in plant and machinery (excl. land & building and other items as per MSMED Act 2006) does not exceed Rs. 25 Lakh, irrespective of the location of the unit.

(b) Small (manufacturing) Enterprises

Enterprises engaged in the manufacture/production, processing or preservation of goods and whose original cost of investment in plant and machinery (excl. land and building and other items as per the Act) is more than Rs. 25 Lakh but does not exceed Rs. 5 crore, irrespective of the location of the unit.

ii) Service Enterprises

(a) Micro (Service) Enterprises

Enterprises engaged in providing/rendering of services and whose original cost of investment in equipment (excl. land and building and furniture, fittings and other items not directly related to the service rendered) does not exceed Rs. 10 Lakh, irrespective of the location of the unit.

b) Small (service) Enterprises

Enterprises engaged in providing/rendering of services and whose original cost of investment in equipment (excl. land and building and furniture, fittings and such items) is more than Rs. 10 Lakh but do not exceed Rs. 2 crore, irrespective of the location of the unit.

c) Other activities

The small and micro (service) enterprises also include small road & water transport operators, small business, professional & self-employed persons, and other service enterprises engaged in activities as listed in the Act and satisfy the investment in equipment criteria i.e. not exceeding Rs. 10 Lakh and Rs. 2 crore respectively.

Loans granted by RRBs to micro and small enterprises (MSE) (manufacturing and services) are eligible for classification under priority sector, provided such enterprises satisfy the definition of MSE sector as contained in MSMED Act, 2006, irrespective of whether the borrowing entity is engaged in export or otherwise.

All advances granted to units in the KVI sector, irrespective of their size of operations, location and amount of original investment in plant and machinery.

Indirect Finance

Indirect finance to small enterprises shall include finance to any person providing inputs to or marketing the output of artisans, village and cottage industries, handlooms and to cooperatives of producers in this sector.

3. Micro Credit

- Loans of very small amount not exceeding Rs. 50,000 per borrower, either directly or indirectly through a SHG/JLG mechanism.
- Loans to distressed persons (other than farmers) to prepay their debt to non-institutional lenders, against appropriate collateral or group security.

4. State sponsored organization for SC/ST

Advances sanctioned to State Sponsored Organisations for Scheduled Castes/ Scheduled Tribes for the specific purpose of purchase and supply of inputs to and/ or the marketing of the outputs of the beneficiaries of these organisations.

5. Education

Loans granted to individuals for educational purposes up to Rs. 10 lakh for studies in India and Rs. 20 lakh for studies abroad. Loans granted to educational institutions which qualify under MSMED Act 2006 are eligible to be classified under priority sector under MSE (services) provided they qualify under MSMED Act 2006.

6. Housing

- Loans up to Rs. 25 lakh, irrespective of location, to individuals for purchase/construction of a dwelling unit per family, excluding loans granted by banks to their own employees, w.e.f. 01.04.2011.
- Loans given for repairs to the damaged dwelling units of families up to Rs. 1 lakh in rural and semi-urban areas and up to Rs. 2 lakh in urban and metropolitan areas.
- Assistance given to any governmental agency for construction of dwelling units or for slum clearance and rehabilitation of slum dwellers, subject to a ceiling of Rs. 5 lakh of loan amount per dwelling unit.
- Assistance given to a non-governmental agency approved by the NHB for the purpose of refinance for construction/reconstruction of dwelling units or for slum clearance and rehabilitation of slum dwellers, subject to a ceiling of loan component of Rs. 10 Lakh per dwelling unit.

7. Weaker Sections

It covers:

- Small and marginal farmers with land holding of 5 acres and less, and landless labourers, tenant farmers and share croppers;
- Artisans, village and cottage industries where individual credit limits do not exceed Rs. 50,000
- SC and ST and DRI beneficiaries.
- Beneficiaries under Swarna Jayanti Shahari Rozgar Yojana (SJSRY), Swarnjayanti Gram Swarozgar Yojana (SGSY) now National Rural Livelihood Mission (NRLM), and Beneficiaries under the Scheme for Rehabilitation of Manual Scavengers (SRMS)
- Loans to distressed poor to repay their debt to informal sector against appropriate collateral, etc.

5.4.3 Guidelines

- Govt. Agencies should complete applications in respect of govt. sponsored schemes like NRLM, etc. and in other areas the bank staff should help the borrowers for this purpose.
- Banks should give acknowledgement for loan applications received from weaker sections.
- All loan applications up to a credit limit of Rs. 25,000 should be disposed of within a fortnight and those for over Rs. 25,000, within 8 to 9 weeks.
- All loan applications for Micro and Small Enterprises up to a credit limit of Rs. 25,000 should be disposed of within 2 weeks and those up to Rs. 5 lakh

within 4 weeks, provided the loan applications are complete in all respects and are accompanied by a 'check list'.

- Branch Managers may reject applications (except in respect of SC/ST) provided the cases of rejection are verified subsequently by the Divisional/ Regional Managers. In the case of proposals from SC/ST, rejection should be at a level higher than that of Branch Manager.
- A register should be maintained at the branch, wherein the date of receipt, sanction/rejection/disbursement with reasons therefore, etc., should be recorded. The register should be made available to all inspecting agencies.
- Banks may disburse all loans for agricultural purposes in cash which will facilitate dealer choice to borrowers and foster an environment of trust. However, banks may continue the practice of obtaining receipts from borrowers.
- Repayment schedule should be based on surplus available, sustenance requirements, life of asset, etc and not on "ad hoc" manner. For people affected by natural calamities, restructuring of the loans may be considered as per NABARD guidelines.
- Rate of interest on various categories of priority sector advances should be as per the RBI directives being issued from time to time.
- No service charges/inspection charges for loans upto Rs. 25,000/- and above the limit, banks may charge as approved by their respective Boards.
- Banks may waive insurance for all categories of priority sector advances upto and inclusive of Rs. 10,000/- and for Micro & Small Enterprises upto Rs. 25,000/-. However, insurance on vehicles and other equipments wherever it is compulsory as per Law or for availing refinance, etc. must be insisted upon.
- Banks should make arrangements for taking photographs of priority sector borrowers wherever it is required at the cost should also be borne by the banks.
- All Branch Managers should be vested with adequate powers to sanction proposals from weaker sections, if not, should exist at district level so as to expedite the sanctions/decisions promptly.
- Banks should put in place a suitable machinery to look into complaints, if any, being received from borrowers and the name and address of the officer with whom complaints can be lodged should also be displayed on the Notice Board of every branch.

6

Introduction to Higher Agricultural Financing Institutions

6.1 RBI

6.1.1 Provision of Finance

- RBI extends short, medium and long-term credits to agriculture through co-operative
- channels. The bulk of the credit granted by the RBI related to short-term to meet Seasonal
- Agricultural Operations (SAO). The RBI Act was amended in 1955 to provide for the
- establishment of two funds, viz., National Agricultural credit (Long Term Operations) Fund and
- National Agricultural Credit (Stabilization) Fund. The three components of medium-term credit are:
  - loans for purchasing shares in co-operative processing societies,
  - loans for agricultural and other allied purposes, including animal husbandry and pisciculture
  - Conversion of short-term agricultural loans into medium term loans when repayment becomes difficult due to natural calamities.

While the first two ((a) and (b)) are financed out of the NAC (LTO) fund and the (c) (Last one) is out of NAC (stabilization) fund. The RBI provides long-term credit as loans to the state governments for contribution to the share capital of the co-operative credit institutions and to NABARD. The RRBs get refinance facility from the RBI up to 50 per cent of their out standing advances.

6.1.2 Promotional Activities

Under rehabilitation programme, most of the cooperative banks have attained viable status. The Study Teams appointed by the RBI have given constructive suggestions to reorganize the co-operative structure on sound lines. The following need special mention: the Committee on Co-operative Land Development Banks (1974), the Committee on Integration of Co-Operative Credit Institutions (1976) committee to study the Interest Rates Spreads in Agricultural Lending Sector etc.

RBI introduced production-oriented credit by means of crop loan system. It has evolved norms to extend medium term finance. It has been issuing appropriate guidelines to Land Development Banks for purposeful and productive utilization of resources. The benefit of deposit insurance cover has been extended to the co-operative banks operative banks so as to facilitate mobilization of deposits.

The RBI evolved a scheme to finance the weak PACS by commercial banks. It initiated the Lead Bank Scheme in 1969 and its impact is seen in branch expansion and other activities of the commercial banks. It advised the banks to participate effectively in the Integrated Rural Development Programme and extend loans to the weaker sections.

6.1.3 Regulatory Functions

As a lender, the RBI not only concerned itself with the quantity of credit but also attempted to improve the quality of credit extended and also the efficiency of the channels through which it is provided to the rural sector. The cooperative banks were brought under certain provisions of the Banking Regulation Act and of the RBI Act in 1966. This enabled the RBI to effectively supervise the co-operative banks as it does over the commercial banks. Under the Credit Authorization Scheme (1976), the co-operative banks should get prior authorization from RBI for providing finance beyond a limit. The RBI frames the overall credit policy on the basis of the credit needs of agriculture. Limits to credit institutions are fixed by taking into account the demand for credit and not just arbitrarily.

The cash / liquidity ratios applicable to co-operatives are lower than those fixed for commercial banks. The co-operatives are enabled to borrow from the RBI at an interest rate which is three per cent below the bank rate in respect of crop loans. They are also permitted to pay slightly higher rates of interest on their deposits. The refinancing functions of RBI relation to rural credit were function of RBI relation to rural credit was taken over by NABARD after its formation in 1982.

6.1.4 Promoting Agricultural Finance

The Reserve Bank of India in a developing economy like ours may be regarded as an engine of growth. It not only regulates bank finance, but deliberately promotes development finance. It has made special efforts in catering to the growing financial needs of agriculture, industry and export sectors of the country.

Agriculture is the king-pin of India's rural economy. Thus, rural credit — agricultural finance — is the prerequisite of agricultural growth and development. Since the inception of planning in our country, the Reserve Bank of India has been paying specific attention to promoting rural/agricultural finance. In fact, the Reserve Bank of India Act, 1934 did assign to the Reserve Bank the responsibility of developing an institutional credit system for the agricultural sector in the country. As such, the Agricultural Credit Department of the Bank was constituted along with the establishment of the Reserve Bank in April 1935, whose main task was to develop cooperative credit movement in agricultural finance.

6.1.5 Funds of RBI

RBI has created two funds for agricultural as recommended by Rural Survey Committee as

(1) National Agricultural Credit (Stabilization Fund)

National Agricultural Credit (Stabilization Fund) which was established in 1956 with a capital of rs. 1 (One) core for medium term purposes. The banks have been contributing rs. 1 crore every year to this fund since its inception .But in view of the widespread scarcity. Contribution and the likely consequent demand on it, it contributed Rs. 6 Crore to this fund on June 30, 1966. One important development that took place during recent past was that is sanctioned loans from it for the first time since its inception.

(2) National Agricultural Credit (Long Term Operation Fund)

National Agricultural Credit (Long Term Operation Fund) was set up with a capital of Rs. 5 crore . The total amount standing to the credit of this fund as on 30 June 1966 was 115 crores. The banks made an enhanced contribution of Rs. 15 crore to this fund in 1967 as compared to rs. 14 crores in the previous year. With a view to accelerate the pace of implementation of the "Crop Loan System "the Reserve Bank published a manual on short term and medium term loans for agricultural purposes in June 1966 .The policy of the bank has been provided agricultural finance mainly through co-operative societies as recommended by the Rural Credit Survey and in conformity with the policy laid down in our five year plans.

6.1.6 Contribution of RBI

Agricultural development is regarded as a prerequisite of economic development of the country. The Reserve Bank of India realizes the following basic contributions of the Agricultural sector in the overall economic development.

- Product contribution making available food and raw material.
- Market contribution providing the market for producer goods and consumer goods produced in the none agricultural sector.

- Factor contribution making available labour and capital to non-agricultural sector, and
- Foreign exchange contribution.

Being the largest industry in the country agriculture is the source of livelihood for over 70% of population in the country. On recognizing the fact that Agriculture is the foundation on which the entire super structure of the growth of industrial and other sectors of the economy has to stand, the RBI develops the Agricultural sector in the following ways :

1. Agricultural Credit Department

According to Section 54 of the RBI Act, it is required to set up a separate Agricultural Credit Department. With the formation of NABARD in 1982, all the activities of this Department have been transferred to NABARD. However, the Rural Planning and Credit Department in the Reserve Bank deals with the following agriculture related matters.

- To study and identify all issues relating to Agricultural credit for policy formulations.
- To provide necessary advice to the Central and State Governments, NABARD, Commercial Banks and Co-operative Banks, in the area of Agricultural Credit and priority Sector Credit.
- To co-ordinate with NABARD the activities of the banking sector relating to agricultural promotion.
- To prepare suitable plans to improve agricultural development in the country.
- To help in the establishment of nationwide warehouse facilities for processing and market of agriculture produce, etc.
- To implement and monitor special rural schemes announced by Government of India for poverty alleviation and employment generation like IRDP schemes.
- To monitor and guide successful operation of Lead Bank Schemes.
- To prescribe policy measures for the operations of Regional Rural Banks.
- To formulate policy for successful conduct of banking ombudsman Scheme.
- To assist in the formation of local area banks.

2. Financial Assistance to Co-operative Sector

The RBI does not provide agriculture finance directly to the farmers. It provides such financial facilities through NABARD which in turn provides financial assistance to the State Co-operative Banks so as to enable them providing credit to rural farmers.

❖ Short-term loans to State Co-operative Banks:

The short-term loans are provided to the co-operative banks for financing seasonal agricultural operations and marketing of agricultural operations and marketing of agricultural produce at a concessional rate.

It also helps the marketing societies in the form of providing working capital for purchasing, stocking and distribution of fertilizers, seeds, pesticides, etc. RBI makes available liberal finance to NABARD at concessional rate of interest for this purpose. The finance is provided through a General Line of Credit to NABARD.

❖ Medium-Term Credit to Co-operative Banks:

With part of RBI funds, NABARD extends medium term loans for the period up to 5 years to the State Co-operative Banks for financing for land improvements, land reclamation, purchase of machinery, construction of small irrigation works, dairy farming, etc.

If natural calamities such as earthquake, floods, fire, famine etc., occur, the farmers who availed short-term loans from the co-operative banks cannot repay their loans in time. Under such circumstances, the RBI permits converting short term loans into medium-term loans. These loans are sanctioned at concessional rates.

3. Establishment of Agricultural Credit Board

In order to co-ordinate the activities of co-operative credit institutions and policies and the functions of RBI, the RBI established a permanent consultative committee in 1951. On the recommendations of the "Rural Credit Survey Committee" it was reorganized in 1956. In 1965 the committee was renamed as "Rural Co-operative Credit Consultative Committee".

Again, in 1970 it was reorganized and renamed as "Agricultural Credit Board". The Governor of the RBI used to be the Chairman of this Board. The Board supervised and coordinated the activities of the RBI and other agencies and ensured the adequate and timely farm credit.

4. Establishment of NABARD

The Reserve Bank of India, in consultation with the Government of India, set up a Committee to Review Arrangement for Institutional Credit for Agriculture and Rural Development (CRAFICARD) in 1979 under the chairmanship of Shri. B. Shivraman.

The CRAFICARD recommended that the NABARD be set-up. The objective was that such a national bank would be able to give undivided attention to the problems of providing production and investment credit to different rural sectors like agriculture, small scale and cottage industries and allied activities in an integrated manner. Based on this recommendation, the Government of India had setup NABARD in July 1982.

Ever since NABARD began functioning, the different functions of the Reserve Bank of India relating to agricultural finance are now performed by NABARD. The entire refinancing work for rural credit has now been transferred to NABARD. The Agricultural Refinance and Development Corporation (ARDC) have been merged with the NABARD.

The National Agricultural Credit (long-term Corporation) Fund and National Agricultural Credit (Stabilization) Fund have been transferred to NABARD.

These funds are now called National Rural Credit (Long term Operations) Fund and National Rural Credit (stabilization) Funds. Central and State Governments also contribute to the funds. The main function of NABARD is discussed below:

NABARD provides refinance to State Co-operative Banks, Regional Rural Banks, State Land Development Banks, State Government and Commercial Banks as indicated below:

Purposes

- Refinance for investment in agriculture and allied activities such as minor irrigation, land development, soil conservation, dairy, sheep, poultry, piggery farm, mechanization, plantation/horticulture, forestry, fishery, storage and market yards, agricultural aviation, biogas and other alternative sources of energy, sericulture, agriculture, animals and animal driven carts, agro-processing, agro services centres, compost plants, modern abattoirs, pump sets, energisation, etc.
- Refinance for artisans/small-scale industries/tiny sector industries, village and cottage industries, handicrafts, etc. (Non-farm Sector).
- Loans to State Governments for share capital contribution to co-operative institutions.
- Investment in share capital/securities of institutions concerned with agriculture and rural development. Period: Available up to a maximum of 25 years.
  - Conversion and Rescheduling Facilities

NABARD provides refinance to eligible institutions normal SCBs and RRBs for conversion and rescheduling of loans under conditions of drought, famine or other natural calamities, military operations, enemy action, etc. Similar facilities are also available in respect of loans made to artisans, small-scale industries, etc.

- Financing Cottage/Villages/Small Scale Industries, etc.

All such industries located in 'rural areas' will be eligible for refinance from NABARD.

NABARD provides refinance for various purposes like minor irrigation, land development, farm mechanization, plantation / horticulture poultry / sheep / piggery, fisheries, dairy development, storage and marketing yields, forestry, goober gas plants, non-farm sector, IRDP, etc. About 80-82 per cent of disbursements are made for farm sector.

Loans provided to small and marginal farmers and other weaker sections of the society enjoy major share of refinance support from NABARD, nearly 80 per cent of the refinance disbursed for the purposes which are relevant for rural households. NABARD also encourages scheduled castes/tribes corporations to formulate special action plan outside. IRDP for SC/ST beneficiaries for which refinance is given under SC/ST Action Plans.

5. Assistance to Co-operative Banks in SFDA and MFAL

Small Farmers Development Agency (SFDA) and Marginal Farmers and Agricultural Labourers (MFAL): To bridge the gap between the small and big farmers, which is likely to be widened in the course of induced development, the All India Rural Credit Review Committee recommended the setting up two agencies, namely, the SFDA and the MFAL, in 1970-71.

The SFDA assists small farmers in the matter of irrigation facilities; land improvement, technical knowhow with a view to changing the farms into progressively prosperous and viable units.

Those holding between 1.25 and 2.5 acres of wet lands or 2.5 and 5 acres of garden lands are described as small farmers. Marginal farmers are those having 1.25 acres of wet lands or 2.5 acres of garden lands. Agricultural labourers comprise those having a homestead and earning 50 per cent or more of their income from agricultural labour wages.

The scheme aims at enabling these farmers to maximize production on their small holdings and to take up subsidiary occupations like poultry and sheep breeding which generate additional income. Credit will be channelized through co-operative institutions at the apex and at the primary levels. The scheme also envisages the development of co-operative marketing by linking up cultivation credit with marketing. The NABARD used to help the co-operative banks in the project of MFAL and SFDA. By this assistance it fulfills the credit needs of economically weak farmers.

6. Reform Measures for Regional Rural Banks

The Regional Rural Banks have come a long way since their inception in 1975 and have now become an integral part of the rural financial system. As a part of financial sector reforms, the RBI along with Government of India and NABARD initiated various measures since 1994 for improving the performance and operations of Regional Rural Banks.

The noteworthy measures among them are: Removal of the condition relating to 100 per cent of landings to 'target group'; providing investment avenues like bonds, share, units of Mutual Funds, etc. Permission to undertake fee-based services like guarantee issuance, locker-room facility, issue of drafts, etc.

The RBI subscribed to the share capital of Central and State Warehousing Corporations. This enables the corporations to establish warehouses at various places of the country. The farmers can avail loan facilities from the banks against warehouse receipts.

7. Facilities to Agriculture

Besides the above specific facilities, the RBI assists the Government of India and NABARD in the formulation and implementation of various schemes meant for agriculture and rural development. Such schemes include the following:

- Intensive agricultural district programme
- Intensive agricultural area programme
- High yielding varieties programme
- Integrated rural development programme
- Community development programme
- Drought prone area programme
- Integrated dry land agriculture development programme, etc.

8. National Agricultural Insurance Scheme

Government of India in July 1999 introduced an Agricultural Insurance Scheme. Farmers availing of seasonal agricultural operations, loans from financial institutions/banks will be covered under the scheme compulsorily. Under the scheme risk insurance will be provided by the Government to cover farmers yield losses arising on their crops due to non- preventable risks like:

- Natural fire and lightening
- Storm, cyclone, etc.
- Flood, landslide
- Drought, dry spells and
- pest/diseases, etc.

The scheme covers crops like sugarcane, potato, cotton and even pulses. The scheme will be implemented through General Insurance Corporation of India (GIC) in co-ordination with commercial banks, co-operative banks and regional rural banks.

The significance of agriculture in India arises also from the fact that the development of agriculture is an essential condition for the development of the national economy. The planned programme for agriculture in India is of recent origin. Because of the vital role played by RBI and subsequently by NABARD, the following benefits are obtained:

There is a steady increase in the area under cultivation. There is a steady rise in average yield per hectare, or rise in agricultural productivity and as a result of an increase in the area as well as increase in yield per hectare, total production of all crops recorded a rising trend.

It is due to the continued institutional credit made available to the agricultural sector first by RBI and later by NABARD and RBI, the food grain production today crossed 200 million tonnes mark. It is estimated to be 202.5 million tonnes in 1998-99.

9. Rural Credit

As of March 2002 more than 1, 40,000 outlets for dispensing rural credit. The highest number of agency is that of the co-op bank followed by Commercial Bank (CBs). Loans disbursed by co-op. institutions towards rural credit has gone up by 42 per cent from Rs. 14,085 crore during 1997-98 to Rs. 27,080 crore during 2001-02.

However, the credit purveyed by commercial banks for agriculture and rural development was the highest at Rs. 31,964 crore during 2001-02. As regards deposits mobilized by banks from rural areas, commercial banks topped the list with 54 per cent of total deposits.

The Reserve Bank does not provide direct finance for industrial activity. The financing of industrial activity is done by financial institutions and banks. The Reserve Bank, however, provided refinance and financial assistance to certain financial institutions and to commercial banks' industrial landings. Presently, RBI provides refinance only towards export credit made by banks.

The Reserve Bank of India, however, plays a vital role in the field of industrial finance. The economic development in modern times has come to be associated with industrialization. The industrial growth largely depends upon the availability of financial facilities.

The money market, mainly for short-term finance provides working capital to industry and commerce, whereas, the capital market primarily for long-term finance provides block or fixed capital to business. These two markets are collectively called as market of credit. The RBI has associated with the Government of India for the establishment of specialized financial institutions as an integral part of the capital market.

The commercial banks were neither willing nor competent to provide long-term loans to Indian Industries as bulk of their funds were in the form of short-term or demand deposits. Thus, RBI realized the need for setting up financial institutions which can specialize in offering long-term financial assistance to industry.

(a) Establishment of Institutional Framework for Industrial Finance. The RBI has played an active role in the field of Industrial Finance its most notable contribution has been in establishing a broad institutional framework to cater to the medium and long-term needs of finance to industrial sector. Some of the institutions in this category for the establishment of which the RBI was responsible are as follows:

- Industrial Finance Corporation
- Industrial Credit and Investment Corporation of India
- Industrial Development Bank of India
- State Finance Corporations
- National Industrial Development Corporation
- State Industrial Development and Investment Corporation
- National Small Industries Corporation
- Unit Trust of India
- Industrial Reconstruction Corporation of India
- Industrial Refinance Corporation of India
- Credit Guarantee Scheme through Export Credit Guarantee

(b) Transfer of Loan Accounts from one Bank to Another. The RBI had laid down certain procedural guidelines/safeguards relating to the transfer of loan accounts by borrowers from one bank to another. These were issued in June, 1977 in pursuance of the recommendations contained in the reports of the Committee on Transfer of Loan Accounts appointed by the Bank.

(c) Bridge Loan. Companies approach banks for provision of bridge loans so as to expedite ongoing capital work, pending disbursal of financial assistance from All India Financial Institutions or against expected equity issues.

In November 1975, banks were advised that bridge loans could be granted to companies without prior authorization of RBI, against the committed financial assistance from the All India Financial Institutions.

However, subsequent to the submission of the Report of the Committee to review the working of the Credit Authorisation Scheme (Marathe Committee)in March 1985, it was made compulsory

(d) Prompt Settlement of Bills of Small Scale Industrial Units. With a view to ensuring prompt settlement of dues of small scale industrial units as also encouraging bills culture, banks have been advised to ensure with effect from January 1, 1998, that the corporate borrowers finance their domestic credit purchases from small scale industrial units, at least to the extent of 25 per cent by way of acceptance of bills drawn on them by the suppliers.

Banks have been advised to monitor the compliance with this requirement through the information system designed by them and charge penal interest in cases of non-compliance.

(e) Role of RBI in Rehabilitation of Sick Industrial Units. The RBI co-ordinates the effects of banks, financial institutions and government agencies in the rehabilitation of such units.

A single window concept for lending has been introduced and made mandatory by the bank in respect of fresh disbursement under working capital as also for disbursement of rehabilitation term loan and subsequent disbursement of working capital under the sanctioned credit limits to facilitate easy and quick disbursal of funds to sick/ weak industrial units at a single source.

(f) High Level Committee on Credit to SSI: The one-man committee on SSI headed by Shri S.L.Kapur, appointed by RBI to review the credit delivery system to SSIs and suggest measures for its improvement. The Committee has made in all 126 recommendations covering wide range of areas pertaining to financing of SSI sector.

The RBI examined these recommendations and advised banks to take necessary steps for immediate implementation. The more important among these are:

- Delegation of more powers to branch managers to grant ado limits.
- Simplification of application forms
- Freedom to banks to decide their own norms for assessment of credit requirements
- Opening of more specialized SSI branches
- Enhancement in the limit for composite loans to Rs.5 lakh
- Strengthening the recovery mechanism.

(g) Export Finance: One of the important areas of industrial finance is Export Credit. Export finance is a preferred finance for the Central Bank and the country because this sector earns valuable foreign exchange. Hence, the RBI continues to prescribe a preferential rate of interest on export finance.

Not only a reduced rate of interest but also exporters avail of a number of benefits against their exports like tax exemption, facility to maintain foreign currency accounts, etc. The RBI also provides refinance facility to commercial banks against their outstanding export credit. The RBI itself does not extend any credit facility to exporters but provide indirect incentives and benefits like cheap credit, adequate credit, refinance facilities, foreign exchange accounts, etc. As per RBI guidelines commercial banks have to extend export credit facilities to all genuine exporters. It is normally granted on the strength of Letter of credit received from overseas buyers.

Types of Credit

The credit is granted fewer than two stages.

(i) Reshipment Credit: Before shipment of goods for the purposes of procurement of raw materials, purchasing them and arranging for the transportation of goods. This type of facility is known as Reshipment finance or packing credit. It is normally extended for a maximum period of 180 days at concessional rate.

The credit can be availed of by exporters either in India Rupee or in foreign currency. However, repayment for foreign currency credit should be made in foreign currency out of export sale.

(ii) Post-Shipment Credit. This credit is extended by banks after shipment of goods. It is in the form of discounting or purchasing the bills drawn by the exporters on overseas buyers. When a customer avails of Preshipment Credit, automatically post-shipment credit will be utilised to repay the earlier credit.

Under post-shipment credit, the RBI prescribes a lower rate of interest for earliest period of export realisations-usually within 90 days of shipment. All export sales should be realised as per RBI guidelines, within 180 days of shipment.

The rate of interest on post-shipment credit increases as the period of realisation becomes longer. It is because the RBI desires to encourage those exporters who will bring in foreign exchange at the earliest.

Foreign Exchange Benefits

The RBI allows exporters to retain a part of (usually 50 per cent) the foreign exchange earned by exporters in our account known as EEFC A/c. (Exchange Earners' Foreign Currency Account). It allows commercial banks to sell foreign exchange to exporters for various businesses including for foreign tours.

Priority Sector

Foreign banks' advances to export sector are considered as priority sector advances. Banks are directed not to reject any export credit proposal unless such decision is taken at higher levels in the bank. The lowest rate of interest charged on export credit was 9 per cent. The Reserve Bank, many a times, changes the rate depending upon general level of interest rate and export performance.

6.2 NABARD

6.2.1 Introduction

The most important development in the field of rural credit has been the setting up of the National Bank for Agriculture and Rural Development (NABARD) in July 1982. It took over from Reserve Bank of India all the functions that the latter performed in the field of rural credit. NABARD is now the open bank for rural credit.

6.2.2 Functions

The main functions of NABARD are as follows:

- It works as an open body to look after the credit requirement of the rural sector.
- It has authority to oversee the functioning of 'the cooperative sector through its Agricultural Credit Department.
- It provides short-term credit (up to 18 months) to State Cooperative Banks for seasonal agricultural operation (crop loans), marketing of crops, purchase and distribution of fertilizers and working capital requirements of cooperative sugar factories.
- It provides medium-term credit (18 months to 7 years) to State Co-operative Banks and RRBs for agricultural purposes purchase of shares of processing societies and conversion of short- term crop loans into medium term loans in areas affected by natural calamities.
- It provides medium and long-term credit (not exceeding 25 years) for investment in agriculture under schematic lending to State Cooperative Banks, Land Development Banks, RRBs and commercial banks.
- It provides long-term assistance in the form of loans to state governments (not exceeding 20 years) for contribution to share capital of cooperative credit institutions.
- It has been entrusted with the responsibility of inspecting District and State Cooperative Banks and RRBs. The inspection of State Land Development Banks and other Federation Cooperative are undertaken on a voluntary basis.
- It maintains a research and development fund to be used to promote research in agriculture and rural development so that projects and programmes can be formulated and designed to suit the requirement of different areas.

6.2.3 NABARD and Rural Credit

It is an apex institution in the field of rural credit. Therefore it does not deal directly with farmers and other rural people. It grants credit to them through the cooperative banks, commercial banks, RRBs.

(1) NABARD provides two types of refinance. The first is extended to RRBs, and apex institutions, namely StCBs and State governments. The other type of refinance is extended to augment resources for ground level deployment of rural credit.

During 2000-04, the NABARD's refinance policy on short term SAO (Seasonal Agricultural Operations) for co-operative banks and RRBs laid emphasis on augmentation of the ground level credit flow through adoption of region- specific strategies and rationalisation of lending policies and procedure.

(2) Rural Infrastructure Development Fund (RIDF) was established in 1995-96 with a corpus of Rs 2000 crore with the major objective of providing funds to state governments and state- owned corporations to enable them to complete various types of rural infrastructure projects.

Loans under RIDF are given for various purposes like irrigation projects, watershed management, construction of rural roads and bridges etc.

(3) The access to credit for the poor from conventional banking is often constrained by lack of collaterals, information asymmetry and high transaction cost associated with small borrowed accounts. Micro finance has emerged as a liable alternative to reach the hitherto reached for their social and economic empowerment through social and financial intermediation, it involves provision of thrift, credit and other financial services and products of very small amounts to the poor for enabling them to raise their income levels and thereby improve living standards.

In operational terms, micro credit involves small loans, up to Rs 25,000, extended to the poor without any collateral for undertaking self-employment project. Such loans are provided through Micro Finance Institutions (MFIs). One of the most popular models of MFI has been the Grameen Bank model, developed originally in Bangladesh and replicated in various parts of the world. Under this model, Non-Government Organisations (NGOs) form and develop self- help groups (SHGs) and provide credit to them.

(4) Kissan Credit Scheme was established in 1998- 99 to facilitate short-term credit to farmers.

(5) Credit Monitoring Arrangement is established with a view to providing to operative banks with more freedom and discretion to operate in an increasingly liberalised and competitive banking environment. NABARD, start in consultation with the Reserve Bank, decided to start the Credit Authorisation Scheme (CAS) with the Credit Monitoring Arrangement (CMA) with effect from the year 2000-2001.

(6) Cooperative Development Fund (CDF) was set up in 1993 with the objective of strengthening the cooperative credit institutions in the areas of organisational structure, human resource development, resource mobilisation, recovery position etc. The assistance is provided to StCBs/SCARDBs/ CCBs)/PCARDBs by way of grant or loan or both.

The following Table shows the contribution of these different sources to the total agricultural credit in India since 1951-52 to 1996.

Table 1: Borrowing of Cultivators from Different Sources (Percentages)

Sources	1951 -52	1961 -62	1971	1981	1995 -96
Non-institutional					
1. Moneylenders	69.7	49.2	36.1	16.1	7.0
2. Traders	5.5	8.8	8.4	3.2	5.0
3. Relatives and friends	14.2	8.8	13.1	8.7	3.0
4. Landlords and others	3.3	14.5	10.7	8.8	10.0
Sub-total (1 to 4)	92.7	81.3	68.3	36.8	25.0
Institutional					
1. Government	3.1	15.5	7.1	3.9	5.0
2. Cooperatives	3.3	2.6	22.0	29.9	40.0
3. Commercial and Rural Banks	0.9	0.6	2.6	29.4	30.0
Sub-total (1 to 3)	7.3	18.7	31.7	63.2	75.0
TOTAL	100.0	100.0	100.0	100.0	100.0

It can be revealed from the above table that among all the different non-institutional sources the contribution of moneylenders was highest and that was to the extent of 69.7%. However, its contribution gradually came down to 49.2% in 1961-62 and then to 7% in 1996. Total contribution of non-institutional source towards agricultural credit has gradually declined from 92.7% in 1951-52 to 25% in 1996.

The share of institutional sources to the total agricultural credit which was 7.3% in 1951-52 gradually increased' to 18.7% in 1961-62 and then to 75% in 1996. Out of these institutional sources cooperatives contributed 40% and commercial banks contributed 30% of the total farm credit in 1996. Although the share of non-institutional sources in the rural areas decreased but still remained very important in supplying credit to the farmers.

The most important development in the field of rural credit is. the setting up of NABARD in July, 1982. This is an apex bank which coordinates the functioning of different financial institutions working for the expansion of the rural credit. It is run by a Board of Directors headed by a chairman.

So far as the supply of credit to agriculture and to rural industries is concerned, this bank performs all the functions including short, medium and long-term refinancing that were previously performed by the Reserve Bank of India. The paid up capital of NABARD is wholly subscribed by the Central Government and the RBI.

The NABARD played an important role in solving the problem of rural indebtedness in India. This aspect would be clear if we study the functions of NABARD, and the overall impact of all activities on Indian agriculture.

6.3 ADB

6.3.1 Introduction

Agriculture the world over continues to contribute to economies in various ways. In developing economies, for instance, it is the livelihood of the very many who live in the rural areas. Now-a-days much of the world's agriculture is struggling and starved of the much needed investment. Heads of state and governments meeting at the 2005 world summit at the United Nations stated that they recognize the need for access to financial services, in particular for the poor, including through microfinance and microcredit.

This reflects what must be and increasingly is a concern of development and poverty eradication policy at national and local levels. Agriculture is typically a capital-intensive industry with investments in farmland, building, machinery, equipment, and breeding livestock dominating the asset structure of most types of farms. However many of these investments have been identified to depend on success to appropriate financial services of which the provision of, and access to credit is prominent.

Agricultural credit plays a very important role in the development of the agricultural sector. It can meet a range of needs and can be critical to the success of agriculture. In fact, circumstantial evidence shows that where agriculture has grown more rapidly, institutional credit has expanded more rapidly. Credit does not only serve as a valuable source of liquidity in responding to risk, but also readily available credit has facilitated many of the significant, long-term challenges in the farm sector increasing commercialization, larger farm sizes, fewer farms, greater capital intensity, adoption of new technology, stronger market coordination and others.

6.3.2 Role

Agricultural Development Bank is reorganising to meet the agricultural sector's needs more effectively – stepping up its strategy of specialisation, creating new products, strengthening its branch network. Growth is happening and agricultural loans are advancing at a remarkable pace.

The main object of the Agricultural Development Bank (ADB) was to provide medium- and long-term financial intermediation and related services to the agricultural and agro-allied sectors in order to make agriculture attractive through modernising its operations; create employment, especially rural employment; provide raw materials for local industries; provide foreign exchange; and improve the living standards of those in the sector. Initially, funding was largely provided by government and multilateral development banks, mainly the World Bank, the European Development Fund, and donor agencies such as USAID. This was on-lent to farmers to finance agricultural projects in the areas of poultry, rubber, oil palm production and processing, and fishing. Funds were used for land preparation, importation of plant and equipment, and working capital. Investments resulted in the commercialisation of agricultural products and by-products, and in transforming several small-scale enterprises into medium-scale enterprises, providing the foundation for Ghana's industrial and socio-economic growth. With the eventual abolition of government funding, the bank had to expand its branch network, mobilise additional deposits, and source funds from multilateral development banks, mainly the African Development Fund (AfDF), the African Development Bank (AfDB), and the World Bank.

Foreign exchange losses (due to financial sector reforms), loan defaults, removals of subsidies on agricultural inputs, and production-related losses led to the rapid deterioration of the bank's overall credit portfolio. This resulted in its re-capitalisation in 1990. This, coupled with the lifting of restrictions on banks to lend at least a quarter of their lending portfolio to the agricultural sector increased pressure on the ADB (with its mandate to finance agriculture) for short-, medium- and long-term agricultural investments.

6.3.3 Current Trend

The ADB began to restructure its business model in 2009 due to its systemic operational uncompetitiveness within the dynamic national and international financial environments. Its restructuring was informed by a three-year corporate strategic plan, aimed at ensuring that the ADB's operations were consistent with modern universal banking practices, in order to meet the requirements of increasingly enlightened investors, customers and shareholders. The restructuring entailed re-aligning the bank's departments and business units to enhance customer satisfaction, comply with industry best practices, and achieve profitability. A new management team was also recruited to actualise the strategic plan and create new strategic business units such as corporate, retail, development finance, transaction banking, and technology. Consequently, rolling out new products was consistent with universal banking functions and was responsive to the needs of agricultural investors. The ADB also expanded its branch network in an effort to bring banking services closer to investors, particularly those engaged in agriculture and agro-based industries. Since June 2010, 22 new branches have been opened in largely rural, agriculturally active communities, corridors and commercial centres with potential for agricultural production.

Since restructuring, the growth of the bank's balance sheet has been impressive, with total net credits growing from GHS 372.86 million (EUR 186 M) in 2009 to GHS

576.99 million (EUR 288 M) in 2010. Over the same period, lending to the agricultural sector climbed from GHS 105.3 million to GHS 174.2 million, representing an increase of 64.5%. This translated into an increase in agriculture's share of the total credit portfolio from 24.1% in 2009 to 28.9% in 2010, showing a purposeful channelling of credit to the bank's agricultural portfolio. Such growth reinforces the ADB's position as the country's number one financier, holding over a third of total agriculture sector credits. The bank's success was recognised in 2010 with its nomination as Bank of Year at Investor Agribusiness Awards. The ADB was recognised for its strides in financial intermediation tailored to the agricultural sector, and its experience accumulated in this area. To further consolidate its position as the premier financial institution in agricultural lending, the ADB's total loan approvals for agricultural purposes made in the first half of 2011 increased by nearly 70%, compared with those made in the same period last year. This significant increase in approvals in agriculture-related loans is fulfilling one of the core objectives of the bank's three-year strategic plan – to increase the share of agriculture in its credit portfolio to 40% by the end of 2012.

6.3.4 Key Challenge

The most significant and critical challenge the ADB faces in its quest to finance and support agricultural investors at all scales is the lack of medium- to long-term financing. This has made it very difficult for the ADB to provide the necessary finance to purchase machinery and plants for agro-processing and to provide working capital for ventures with long gestation periods. The lack of term financing has also made it difficult to consistently develop agricultural and agro-processing industries, thus affecting the sustainable growth of the economy of the country.

ADB's Strategy in Agricultural Sector

To support its agricultural sector strategy, the ADB has developed a sharpened focus on agriculture, by making available asset financing facilities to assist in developing each segment of the value chain. Products and lending facilities have been designed and adapted to provide credit to each sector and type of client. These are tailored to clients' needs, using the bank's understanding of the drivers and structures of the underlying businesses, and the business processes involved. Consequently, its product range is large and flexible: facilities can take the form of short-, medium- or long-term loans; overdrafts; and leasing, depending on the nature/type of financing required. Key sub-sectors where the ADB has a presence include industrial crops (raw materials for industry), food crops, export crops, agro-processing, agro-marketing, agribusiness products, agro inputs, and agro services. Financing of certain farming operations could take the form of a five-year loan, at a special interest rate of 15% (while base rate was 16.75%), using a variety of potential securities, such as cash collateral, life policies, landed properties, plant and equipment, and personal guarantors, among others), which could be granted to individuals, farmer-based organisations, groups, and companies. There are other specialised financial products, like institutionalised personal loans for some farmers, like rubber farmers.

6.3.5 Vision and Potential Opportunities

The ADB's vision for the agricultural sector is to help it and its value-chain grow by providing medium- and long-term financing on a strategic and sustained basis. This is premised on the agriculture sector's huge potential, given global market prices for most agricultural products. However, executing this vision and pursuing the available opportunities require the availability of term financing from various sources, which should be accessible to the ADB. Furthermore, with the establishment of micro-finance institutions and the growth of Rural Banks in recent times, the monopoly that the bank used to enjoy in financing small-scale agriculture is gradually being eroded. In spite of significant challenges, the ADB has made giant strides in the past two years. To consolidate these, the bank's strategy for 2011 involves. Though supporting out-growers is in its early stages, the ADB is hopeful because of the partnership it has developed with international development agencies such as AFD.

6.3.6 Microcredit Delivery Policies

Formal institutional credit schemes for Small Scale Enterprises (SSEs) development dated from late 1969, with the establishment of the Credit Guarantee Scheme by Bank to assist entrepreneurs who generally found it difficult to benefit from bank credit. However, enterprises that were covered by the Scheme included those in agriculture, livestock breeding, manufacturing, and mining among others Not withstanding, in 1970, the government set up the Ghana Business Promotion Programme, which became the Ghanaian Enterprise Development Commission in 1975, to administer the small Business Credit Scheme. The objective of the commission was to provide financial and technical assistance to newly established and existing SSEs and Ghanaian Managers of formerly alien-owned manufacturing enterprises. April 1990, the commission and the National Board for Small-Scale industries (NBSSI) were merged under the lather in order to avoid duplication of efforts in SSE promotion.

In addition, the government established a credit line under the programme of Action to Mitigate the Social Cost of Adjustment and Development (PAMSCAD) to assist SSEs in the financial needs. This is part of the measures adopted by government to cushion the hardships brought to some vulnerable groups, especially under its Structural Adjustment Programme (SAP). The credit is being managed by the NBSSI to assist entrepreneurs to procure scarce but essential raw materials.

Furthermore, there are also the rural banks which has been established to help SSEs in their credit needs. These are independent unit banks whose operation seek to bring banking facilities to the doorsteps of rural people, by making institutional credit available to them, mobilizing saving in the rural communities, extending credit to farmers and other rural producers and serving as a catalyst in rural development. This catalogue of credit schemes indicates the laudable intentions of government in helping to promote the development of SSEs in the country.

6.3.7 Credit Delivery Schemes

Agricultural Development Bank (ADB) offers a number of credit delivery schemes. One of these schemes is group credit. A group system is a method of loan delivery in which loans and their repayment are made the responsibility of organized groups rather than individuals. Here, credit resources may be generally attributed through small groups rather than the loans being allocated on an individual by individual basis. Solidarity groups make an efficient use of local knowledge and social relations and are characterized by social cohesion and mutual trust. Therefore, in the absence of enforceable contracts and collateral in microcredit, social 'connectedness' guarantees inter-temporal commitments.

The second scheme is individual credit. Under this scheme, individuals are granted credit and they are solely liable for repayment and default. Many of the programmes operating on this policy have been most successful in areas where good infrastructure coverage is available and marketing has been stimulated. It also made possible in formal financial institutions where beneficiaries are required to produce some form of collateral security.

Supervised credit is another scheme. Credit to borrowers who lack education and ability to keep records, is best administered under close supervision (Aryeetey, 1995). Therefore credit should be granted only after a thorough investigation of activities of the borrowers and may also provide other credit related services such as education and training

6.3.8 Micro-Credit

Micro-credit delivery is a form of financial inter-mediation performed by both banking and non-bank financial institutions. It is used to denote a special kind of contractual agreement that facilitates the transfer of temporal free resources of individuals or organizations to others who have temporal need for them. Thus microcredit refers to any financial or material resources mobilized by the bank or non-bank financial institutions, with the aim of lending to small-scale borrowers. This is usually to assist them get out of their depressed economic circumstances. As seen, micro-credit is a means that enables persons to extend their control as distinct from ownership of a small amount of resources, to temporally meet their socioeconomic objectives. However, for micro-credit schemes to be sustainable and effectively address the needs of the poor especially the skill and education level of beneficiaries need to be considered. Thus a training and educational package should constitute an integral part of the scheme.

Challenges of Accessing Microcredit

The measures of government and other organizations to promote the development of SSEs in both urban and rural areas by making banking facilities close to their doorsteps have failed. This is as a result of poor monetary structures and policies in place. Credit from government or institutional sources often involves delay and completion of many facilities, too complicated for the unsophisticated small-scale and often illiterate local entrepreneurs. In addition, collateral requirements, complex procedures, poor communication and inadequate banking network are all factors which restrict the availability of institutional credit to the majority of households.

6.4 IMF

6.4.1 Introduction

The International Monetary Fund (IMF) is an international organization headquartered in Washington, D.C., consisting of "189 countries working to foster global monetary cooperation, secure financial stability, facilitate international trade, promote high employment and sustainable economic growth, and reduce poverty around the world."

6.4.2 Functions

According to the IMF itself, it works to foster global growth and economic stability by providing policy, advice and financing the members, by working with developing nations to help them achieve macroeconomic stability and reduce poverty. The rationale for this is that private international capital markets function imperfectly and many countries have limited access to financial markets. Such market imperfections, together with balance-of-payments financing, provide the justification for official financing, without which many countries could only correct large external payment imbalances through measures with adverse economic consequences. The IMF provides alternate sources of financing.

Upon the founding of the IMF, its three primary functions were: to oversee the fixed exchange rate arrangements between countries, thus helping national governments manage their exchange rates and allowing these governments to prioritize economic growth, and to provide short-term capital to aid the balance of payments. This assistance was meant to prevent the spread of international economic crises. The IMF was also intended to help mend the pieces of the international economy after the Great Depression and World War II, as well, to provide capital investments for economic growth and projects such as infrastructure.

The IMF's role was fundamentally altered by the floating exchange rates post-1971. It shifted to examining the economic policies of countries with IMF loan agreements to determine if a shortage of capital was due to economic fluctuations or economic policy. The IMF also researched what types of government policy would ensure economic recovery. A particular concern of the IMF was to prevent financial crisis, such as those in Mexico 1982, Brazil in 1987, East Asia in 1997–98 and Russia in 1998, from spreading and threatening the entire global financial and currency system. The challenge was to promote and implement policy that reduced the frequency of crises among the emerging market countries, especially the middle-income countries which are vulnerable to massive capital outflows. Rather than maintaining a position of oversight of only exchange rates, their function became one of surveillance of the overall macroeconomic performance of member countries. Their role became a lot more active because the IMF now manages economic policy rather than just exchange rates.

In addition, the IMF negotiates conditions on lending and loans under their policy of conditionality, which was established in the 1950s. Low-income countries can borrow on concessional terms, which means there is a period of time with no interest rates, through the Extended Credit Facility (ECF), the Standby Credit Facility (SCF) and the Rapid Credit Facility (RCF). Nonconcessional loans, which include interest rates, are provided mainly through Stand-By Arrangements (SBA), the Flexible Credit Line (FCL), the Precautionary and Liquidity Line (PLL), and the Extended Fund Facility. The IMF provides emergency assistance via the Rapid Financing Instrument (RFI) to members facing urgent balance-of-payments needs.

6.4.3 Lending by IMF

A country in severe financial trouble, unable to pay its international bills, poses potential problems for the stability of the international financial system, which the IMF was created to protect. Any member country, whether rich, middle-income, or poor, can turn to the IMF for financing if it has a balance of payments need—that is, if it cannot find sufficient financing on affordable terms in the capital markets to make its international payments and maintain a safe level of reserves.

IMF loans are meant to help member countries tackle balance of payments problems, stabilize their economies, and restore sustainable economic growth. This crisis resolution role is at the core of IMF lending. At the same time, the global financial crisis has highlighted the need for effective global financial safety nets to help countries cope with adverse shocks. A key objective of recent lending reforms has therefore been to complement the traditional crisis resolution role of the IMF with more effective tools for crisis prevention.

The IMF is not a development bank and, unlike the World Bank and other development agencies, it does not finance projects.

Changing nature of lending

About four out of five member countries have used IMF credit at least once. But the amount of loans outstanding and the number of borrowers have fluctuated significantly over time. In the first two decades of the IMF's existence, more than half of its lending went to industrial countries. But since the late 1970s, these countries have been able to meet their financing needs in the capital markets.

The oil shock of the 1970s and the debt crisis of the 1980s led many lower- and lower-middle-income countries to borrow from the IMF. In the 1990s, the transition process in central and eastern Europe and the crises in emerging market economies led to a further increase in the demand for IMF resources.

In 2004, benign economic conditions worldwide meant that many countries began to repay their loans to the IMF. As a consequence, the demand for the Fund's resources dropped off sharply.

But in 2008, the IMF began making loans to countries hit by the global financial crisis The IMF currently has programs with more than 50 countries around the world and has committed more than $325 billion in resources to its member countries since the start of the global financial crisis.

While the financial crisis has sparked renewed demand for IMF financing, the decline in lending that preceded the financial crisis also reflected a need to adapt the IMF's lending instruments to the changing needs of member countries. In response, the IMF conducted a wide-ranging review of its lending facilities and terms on which it provides loans.

In March 2009, the Fund announced a major overhaul of its lending framework, including modernizing conditionality, introducing a new flexible credit line, enhancing the flexibility of the Fund's regular stand-by lending arrangement, doubling access limits on loans, adapting its cost structures for high-access and precautionary lending, and streamlining instruments that were seldom used. It has also speeded up lending procedures and redesigned its Exogenous Shocks Facility to make it easier to access for low-income countries. More reforms have since been undertaken, most recently in November 2011.

Lending to preserve financial stability

Article I of the IMF's Articles of Agreement states that the purpose of lending by the IMF is "...to give confidence to members by making the general resources of the Fund temporarily available to them under adequate safeguards, thus providing them with opportunity to correct maladjustments in their balance of payments without resorting to measures destructive of national or international prosperity."

In practice, the purpose of the IMF's lending has changed dramatically since the organization was created. Over time, the IMF's financial assistance has evolved from helping countries deal with short-term trade fluctuations to supporting adjustment and addressing a wide range of balance of payments problems resulting from terms of trade shocks, natural disasters, post-conflict situations, broad economic transition, poverty reduction and economic development, sovereign debt restructuring, and confidence-driven banking and currency crises. Today, IMF lending serves three main purposes.

- First, it can smooth adjustment to various shocks, helping a member country avoid disruptive economic adjustment or sovereign default, something that would be extremely costly, both for the country itself and possibly for other countries through economic and financial ripple effects (known as contagion).
- Second, IMF programs can help unlock other financing, acting as a catalyst for other lenders. This is because the program can serve as a signal that the country has adopted sound policies, reinforcing policy credibility and increasing investors' confidence.
- Third, IMF lending can help prevent crisis. The experience is clear: capital account crises typically inflict substantial costs on countries themselves and on other countries through contagion. The best way to deal with capital account problems is to nip them in the bud before they develop into a full-blown crisis.

Conditions for lending

When a member country approaches the IMF for financing, it may be in or near a state of economic crisis, with its currency under attack in foreign exchange markets and its international reserves depleted, economic activity stagnant or falling, and a large number of firms and households going bankrupt. In difficult economic times, the IMF helps countries to protect the most vulnerable in a crisis.

The IMF aims to ensure that conditions linked to IMF loan disbursements are focused and adequately tailored to the varying strengths of members' policies and fundamentals. To this end, the IMF discusses with the country the economic policies that may be expected to address the problems most effectively. The IMF and the government agree on a program of policies aimed at achieving specific, quantified goals in support of the overall objectives of the authorities' economic program. For example, the country may commit to fiscal or foreign exchange reserve targets.

The IMF discusses with the country the economic policies that may be expected to address the problems most effectively. The IMF and the government agree on a program of policies aimed at achieving specific, quantified goals in support of the overall objectives of the authorities' economic program. For example, the country may commit to fiscal or foreign exchange reserve targets.

Loans are typically disbursed in a number of installments over the life of the program, with each installment conditional on targets being met. Programs typically last up to 3 years, depending on the nature of the country's problems, but can be followed by another program if needed. The government outlines the details of its economic program in a "letter of intent" to the Managing Director of the IMF. Such letters may be revised if circumstances change.

For countries in crisis, IMF loans usually provide only a small portion of the resources needed to finance their balance of payments. But IMF loans also signal that a country's economic policies are on the right track, which reassures investors and the official community, helping countries find additional financing from other sources.

Main lending facilities

In an economic crisis, countries often need financing to help them overcome their balance of payments problems. Since its creation in June 1952, the IMF's Stand-By Arrangement (SBA) has been used time and again by member countries, it is the IMF's workhorse lending instrument for emerging market countries. Rates are non-concessional, although they are almost always lower than what countries would pay to raise financing from private markets. The SBA was upgraded in 2009 to be more flexible and responsive to member countries' needs. Borrowing limits were doubled with more funds available up front, and conditions were streamlined and simplified. The new framework also enables broader high-access borrowing on a precautionary basis.

The Flexible Credit Line (FCL) is for countries with very strong fundamentals, policies, and track records of policy implementation. It represents a significant shift in how the IMF delivers Fund financial assistance, particularly with recent enhancements, as it has no ongoing conditions and no caps on the size of the credit line. The FCL is a renewable credit line, which at the country's discretion could be for either 1-2 years, with a review of eligibility after the first year. There is the flexibility to either treat the credit line as precautionary or draw on it at any time after the FCL is approved. Once a country qualifies (according to pre-set criteria), it can tap all resources available under the credit line at any time, as disbursements would not be phased and conditioned on particular policies as with traditional IMF-supported programs. This is justified by the very strong track records of countries that qualify to the FCL, which give confidence that their economic policies will remain strong or that corrective measures will be taken in the face of shocks.

The Precautionary and Liquidity Line (PLL) builds on the strengths and broadens the scope of the Precautionary Credit Line (PCL). The PLL provides financing to meet actual or potential balance of payments needs of countries with sound policies, and is intended to serve as insurance and help resolve crises. It combines a qualification process (similar to that for the FCL) with focused ex-post conditionality aimed at addressing vulnerabilities identified during qualification. Its qualification requirements signal the strength of qualifying countries' fundamentals and policies, thus contributing to consolidation of market confidence in the country's policy plans. The PLL is designed to provide liquidity to countries with sound policies under broad circumstances, including countries affected by regional or global economic and financial stress.

The Rapid Financing Instrument (RFI) provides rapid and low-access financial assistance to member countries facing an urgent balance of payments need, without the need for a full-fledged program. It can provide support to meet a broad range of urgent needs, including those arising from commodity price shocks, natural disasters, post-conflict situations and emergencies resulting from fragility.

The Extended Fund Facility is used to help countries address balance of payments difficulties related partly to structural problems that may take longer to correct than macroeconomic imbalances. A program supported by an extended arrangement usually includes measures to improve the way markets and institutions function, such as tax and financial sector reforms, privatization of public enterprises.

The Trade Integration Mechanism allows the IMF to provide loans under one of its facilities to a developing country whose balance of payments is suffering because of multilateral trade liberalization, either because its export earnings decline when it loses preferential access to certain markets or because prices for food imports go up when agricultural subsidies are eliminated.

Lending to low-income countries

To help low-income countries weather the severe impact of the global financial crisis, the IMF has revamped its concessional lending facilities to make them more flexible and meet increasing demand for financial assistance from countries in need. These changes became effective in January 2010. Once additional loan and subsidy resources are mobilized, these changes will boost available resources for low-income countries to $17 billion through 2014. To ensure resources are available for lending to low-income countries beyond 2014, the IMF approved an additional $2.7 billion in remaining windfall profits from gold sales as part of a strategy to make lending to low-income countries sustainable.

Three types of loans were created under the new Poverty Reduction and Growth Trust (PRGT) as part of this broader reform: the Extended Credit Facility, the Rapid Credit Facility and the Standby Credit Facility. The Extended Credit Facility (ECF) provides financial assistance to countries with protracted balance of payments problems. The ECF succeeds the Poverty Reduction and Growth Facility (PRGF) as the Fund's main tool for providing medium-term support LICs, with higher levels of access, more concessional financing terms, more flexible program design features, as well as streamlined and more focused conditionality.

The Rapid Credit Facility (RCF) provides rapid financial assistance with limited conditionality to low-income countries (LICs) facing an urgent balance of payments need. The RCF streamlines the Fund's emergency assistance, provides significantly higher levels of concessionality, can be used flexibly in a wide range of circumstances, and places greater emphasis on the country's poverty reduction and growth objectives.

The Standby Credit Facility (SCF) provides financial assistance to low-income countries (LICs) with short-term balance of payments needs. It provides support under a wide range of circumstances, allows for high access, carries a low interest rate, can be used on a precautionary basis, and places emphasis on countries' poverty reduction and growth objectives.

Several low-income countries have made significant progress in recent years toward economic stability and no longer require IMF financial assistance. But many of these countries still seek the IMF's advice, and the monitoring and endorsement of their economic policies that comes with it. To help these countries, the IMF has created a program for policy support and signaling, called the Policy Support Instrument.

6.5 World bank

6.5.1 Introduction

Since inception in 1944, the World Bank has expanded from a single institution to a closely associated group of five development institutions. Our mission evolved from the International Bank for Reconstruction and Development (IBRD) as facilitator of post-war reconstruction and development to the present day mandate of worldwide poverty alleviation in close coordination with our affiliate, the International Development Association, and other members of the World Bank Group, the International Finance Corporation (IFC), the Multilateral Guarantee Agency (MIGA), and the International Centre for the Settlement of Investment Disputes (ICSID).

Once we had a homogeneous staff of engineers and financial analysts, based solely in Washington, DC. Today, we have a multidisciplinary and diverse staff that includes economists, public policy experts, sector experts and social scientists- and now more than a third of our staff is based in country offices. Reconstruction remains an important part of our work. However, the global challenges in the world compel us to focus on:

- poverty reduction and the sustainable growth in the poorest countries;
- solutions to the special challenges of post-conflict countries and fragile states;
- development solutions with customized services as well as financing for middle-income countries;
- regional and global issues that cross national borders--climate change, infectious diseases, and trade;
- greater development and opportunity in the Arab world;

- pulling together the best global knowledge to support development.

At today's World Bank, poverty reduction through an inclusive and sustainable globalization remains the overarching goal of our work. The World Bank is a vital source of financial and technical assistance to developing countries around the world. Its mission is to fight poverty with passion and professionalism for lasting results and to help people help themselves and their environment by providing resources, sharing knowledge, building capacity and forging partnerships in the public and private sectors. We are not a bank in the common sense; we are made up of two unique development institutions owned by 187 member countries: the International Bank for Reconstruction and Development (IBRD) and the International Development Association (IDA). Each institution plays a different but collaborative role in advancing the vision of inclusive and sustainable globalization. The IBRD aims to reduce poverty in middle-income and creditworthy poorer countries, while IDA focuses on the world's poorest countries.

Their work is complemented by that of the International Finance Corporation (IFC), Multilateral Investment Guarantee Agency (MIGA) and the International Centre for the Settlement of Investment Disputes (ICSID). Together, we provide low-interest loans, interestfree credits and grants to developing countries for a wide array of purposes that include investments in education, health, public administration, infrastructure, financial and private sector development, agriculture and environmental and natural resource management. The World Bank, established in 1944, is headquartered in Washington, D.C. To ensure countries continue to have access to the best global expertise and cutting-edge knowledge, the World Bank Group is revising its programs to assist the poor, as well as its range of financing options, to meet pressing development priorities.

6.5.2 Challenges

The World Bank had made the world's challenge—to reduce global poverty- the bank's challenge. The Bank focuses on achievement of the Millennium Development Goals that call for the elimination of poverty and sustained development. The goals provide the bank with targets and yardsticks for measuring results. The bank's mission is to help developing countries and their people reach the goals by working with our partners to alleviate poverty. It addresses global challenges in ways that advance an inclusive and sustainable globalization—that overcome poverty, enhance growth with care for the environment, and create individual opportunity and hope.

Six strategic themes drive bank's efforts. By focusing on these strategic themes, the Bank delivers technical, financial and other assistance to those most in need and where it can have the greatest impact and promote growth: to the poorest countries, fragile states and the Arab world; to middle-income countries; to solving global public goods issues; and to delivering knowledge and learning services

6.5.3 Operations

The World Bank's two closely affiliated entities—the International Bank for Reconstruction and Development (IBRD) and the International Development Association (IDA)—provide low or no interest loans (credits) and grants to countries that have unfavorable or no access to international credit markets. Unlike other financial institutions, we do not operate for profit. The IBRD is market-based, and we use our high credit rating to pass the low interest we pay for money on to our borrowers—developing countries. We pay for our own operating costs, since we don't look to outside sources to furnish funds for overhead. So, where does the money come from to operate the World Bank, and how do we use the funds to carry out our mission?

Fund Generation

IBRD lending to developing countries is primarily financed by selling AAA-rated bonds in the world's financial markets. While IBRD earns a small margin on this lending, the greater proportion of its income comes from lending out its own capital. This capital consists of reserves built up over the years and money paid in from the Bank's 185 member country shareholders. IBRD's income also pays for World Bank operating expenses and has contributed to IDA and debt relief.

IDA is the world's largest source of interest-free loans and grant assistance to the poorest countries. IDA's funds are replenished every three years by 40 donor countries. Additional funds are regenerated through repayments of loan principal on 35-to-40-year, no-interest loans, which are then available for re-lending. IDA accounts for more than 40% of our lending.

Loans

Through the IBRD and IDA, we offer two basic types of loans and credits: investment operations and development policy operations. Countries use investment operations for goods, works and services in support of economic and social development projects in a broad range of economic and social sectors. Development policy operations (formerly known as adjustment loans) provide quick-disbursing financing to support a country's policy and institutional reforms.

Each borrower's project proposal is assessed to ensure that the project is economically, financially, socially and environmentally sound. During loan negotiations, the Bank and borrower agree on the development objectives, outputs, performance indicators and implementation plan, as well as a loan disbursement schedule. While we supervise the implementation of each loan and evaluate its results, the borrower implements the project or program according to the agreed terms. As more than 30% of our staff is based in over 100 country offices worldwide, three-fourths of outstanding loans are managed by country directors located away from the World Bank offices in Washington.

IDA long term loans (credits) are interest free but do carry a small service charge of 0.75 percent on funds paid out. IDA commitment fees range from zero to 0.5 percent on undisbursed credit balances. For FY09 commitment fees have been set at 0.0 percent.

Trust Funds and Grants

Donor governments and a broad array of private and public institutions make deposits in Trust funds that are housed at the World Bank. These donor resources are leveraged for a broad range of development initiatives. The initiatives vary significantly in size and complexity, ranging from multibillion dollar arrangements—such as Carbon Finance; the Global Environment Facility; the Heavily Indebted Poor Countries Initiative; and the Global Fund to Fight AIDS, Tuberculosis, and Malaria—to much smaller and simpler freestanding ones.

The Bank also mobilizes external resources for IDA concessionary financing and grants, as well as funds for non-lending technical assistance and advisory activities to meet the special needs of developing countries, and for co-financing of projects and programs.

Direct World Bank grants to civil society organizations emphasize broad-based stakeholder participation in development, and aim to strengthen the voice and influence of poor and marginalized groups in the development process. IDA grants—which are either funded directly or managed through partnerships have been used to:

- Relieve the debt burden of heavily indebted poor countries
- Improve sanitation and water supplies
- Support vaccination and immunization programs to reduce the incidence of communicable diseases like malaria
- Combat the HIV/AIDS pandemic
- Support civil society organizations
- Create initiatives to cut the emission of greenhouse gases

Analytic and Advisory Services

While we are best known as a financier, another of our roles is to provide analysis, advice and information to our member countries so they can deliver the lasting economic and social improvements their people need. We do this in various ways. One is through economic research and data collection on broad issues such as the environment, poverty, trade and globalization Another is through country-specific, nonlending activities such as economic and sector work, where we evaluate a country's economic prospects by examining its banking systems and financial markets, as well as trade, infrastructure, poverty and social safety net issues, for example.

We also draw upon the resources of our knowledge bank to educate clients so they can equip themselves to solve their development problems and promote economic growth. By knowledge bank we mean the wealth of contacts, knowledge, information and experience we've acquired over the years, country by country and project by project, in our development work. Our ultimate aim is to encourage the knowledge revolution in developing countries.

These are only some of the ways our analyses, advice and knowledge are made available to our client countries, their government and development professionals, and the public:

- Poverty Assessments
- Public Expenditure Reviews
- Country Economic Reports
- Sector Reports
- Development

Capacity Building

Another core Bank function is to increase the capabilities of our partners, the people in developing countries, and our own staff —to help them acquire the knowledge and skills they need to provide technical assistance, improve government performance and delivery of services, promote economic growth and sustain poverty reduction programs. Linkages to knowledge-sharing networks such as these have been set up by the Bank to address the vast needs for information and dialogue about development:

- Advisory Services and Ask Us help desks make information available by topic via telephone, fax, email and the web. There are more than 25 advisory services at the Bank. Staff members who respond to inquiries add value to the work of clients, partners and our own staff by responding quickly to their knowledge needs. Often, they are the first and possibly the only contact the public at large—especially people in developing countries--have with the World Bank.
- Global Development Learning Network is an extensive network of distance learning centers that uses advanced information and communications technologies to connect people working in development around the world.
- Knowledge for Development offers policy advice to client countries on the four pillars of a knowledge economy: economic and institutional regime, education, innovation, and information and communication technologies (ICTs) to help clients make the transition to a knowledge economy.
- Capacity Development Resource Center is a repository of literature, case studies, lessons learned, and good practices in the area of capacity development, the key to development effectiveness.

- World Bank Institute Global and Regional Programs bring together leading development practitioners online and face-to-face to exchange experiences and to develop skills.
- B-SPAN webcasting service is an Internet-based broadcasting station. The station presents World Bank seminars, workshops and conferences on sustainable development and poverty reduction via streaming video. The unedited discussions and debates about pressing development issues attract government officials, development practitioners, academics, students, researchers, journalists, NGO representatives, and the public-at-large.

7

Insurance and Credit

7.1 Deposit Insurance

7.1.1 Introduction

Importance of deposit insurance has risen significantly in the last two decades. During this period, waves of systemic crises in banking have affected most of the countries around the globe. There have been as many as 112 episodes of systemic crisis in 93 countries and 51 episodes of borderline crisis in 46 countries. Total fiscal costs incurred in the 1997 Thai and Korean banking crises exceeded 30 percent of GDP, and in Indonesia, budgetary costs for the banking crises approached 50 percent of GDP. Notwithstanding the increase in total fiscal costs, full costs of these crises go beyond direct fiscal costs in terms of welfare and loss of GDP.

In India, the issue of depositor protection came to the fore in 1938 on the failure of the Travancore National and Quilon Bank, the largest bank in southern India followed by the banking crisis in Bengal between 1946 and 1948. In the early 1950s, both the Rural Banking Enquiry Committee and the Shroff Committee adverted to the advantages of insuring bank deposits. It was, however, felt that the measures be held in abeyance till the Banking Companies Act, 1949 came into force and comprehensive arrangements were made for the supervision and inspection of banks by the Reserve Bank of India. With the subsequent failures of Laxmi Bank and the Palai Central Bank in 1961, which shook the confidence in the banking system, the Reserve Bank responded quickly with a blue print for insuring bank deposits.

After the passage of the Deposit Insurance Corporation (DIC) Bill in 1961, the Deposit Insurance Act, 1961 came into force on January 1, 1962 and the Deposit Insuran ce Corporation commenced functioning under the aegis of the Reserve Bank of India. Incidentally, the Reserve Bank, in 1971, had also promoted the Credit Guarantee Corporation of India aimed at encouraging commercial banks to cater to the credit needs of hitherto neglected sectors, particularly the weaker sections, by providing guarantee cover to loans and advances granted by credit institutions to small and needy borrowers under priority sector lending. With the integration of the deposit insurance and credit guarantee functions, the Deposit Insurance and Credit Guarantee Corporation (DICGC), as it is presently known, came into being in July 1978. Incidentally, the Corporation is presently not operating credit guarantee schemes, as no credit institution is currently participating in any such schemes.

7.2 Deposit Insurance Scheme in India

The concept of insuring deposits kept with banks received attention for the first time in the year 1948 after the banking crises in Bengal. The question came up for reconsideration in the year 1949, but it was decided to hold it in abeyance till the Reserve Bank of India ensured adequate arrangements for inspection of banks. Subsequently, in the year 1950, the Rural Banking Enquiry Committee also supported the concept. Serious thought to the concept was, however, given by the Reserve Bank of India and the Central Government after the crash of the Palai Central Bank Ltd., and the Laxmi Bank Ltd. in 1960. The Deposit Insurance Corporation (DIC) Bill was introduced in the Parliament on August 21, 1961. After it was passed by the Parliament, the Bill got the assent of the President on December 7, 1961 and the Deposit Insurance Act, 1961 came into force on January 1, 1962.

The Deposit Insurance Scheme was initially extended to functioning commercial banks only. This included the State Bank of India and its subsidiaries, other commercial banks and the branches of the foreign banks operating in India.

Since 1968, with the enactment of the Deposit Insurance Corporation (Amendment) Act, 1968, the Corporation was required to register the 'eligible co-operative banks' as insured banks under the provisions of Section 13 A of the Act. An eligible co-operative bank means a co-operative bank (whether it is a State co-operative bank, a Central co-operative bank or a Primary co-operative bank) in a State which has passed the enabling legislation amending its Co-operative Societies Act, requiring the State Government to vest power in the Reserve Bank to order the Registrar of Co-operative Societies of a State to wind up a co-operative bank or to supersede its Committee of Management and to require the Registrar not to take any action for winding up, amalgamation or reconstruction of a co-operative bank without prior sanction in writing from the Reserve Bank of India.

Further, the Government of India, in consultation with the Reserve Bank of India, introduced a Credit Guarantee Scheme in July 1960. The Reserve Bank of India was entrusted with the administration of the Scheme, as an agent of the Central Government, under Section 17 (11 A)(a) of the Reserve Bank of India Act, 1934 and was designated as the Credit Guarantee Organization (CGO) for guaranteeing the advances granted by banks and other Credit Institutions to small scale industries. The Reserve Bank of India operated the scheme up to March 31, 1981.

The Reserve Bank of India also promoted a public limited company on January 14, 1971, named the Credit Guarantee Corporation of India Ltd. (CGCI). The main thrust of the Credit Guarantee Schemes, introduced by the Credit Guarantee Corporation of India Ltd., was aimed at encouraging the commercial banks to cater to the credit needs of the hitherto neglected sectors, particularly the weaker sections of the society engaged in non-industrial activities, by providing guarantee cover to the loans and advances granted by the credit institutions to small and needy borrowers covered under the priority sector.

With a view to integrating the functions of deposit insurance and credit guarantee, the above two organizations (DIC & CGCI) were merged and the present Deposit Insurance and Credit Guarantee Corporation (DICGC) came into existence on July 15, 1978. Consequently, the title of Deposit Insurance Act, 1961 was changed to 'The Deposit Insurance and Credit Guarantee Corporation Act, 1961 '.

Effective from April 1, 1981, the Corporation extended its guarantee support to credit granted to small scale industries also, after the cancellation of the Government of India's credit guarantee scheme. With effect from April 1, 1989, guarantee cover was extended to the entire priority sector advances, as per the definition of the Reserve Bank of India. However, effective from April 1, 1995, all housing loans have been excluded from the purview of guarantee cover by the Corporation.

7.2.1 Pay box system versus extended mandate systems

The DIS in India is a paybox system. Apart from infusing confidence of the common man in the banking system, a Paybox systems performs three major tasks:

- Contributing to the smooth functioning of the payments system
- Reducing the potential for moral hazard by placing limits on insurance or introducing coinsurance
- Facilitating the exit of problem banks. All these advantages have been experienced in India especially when dealing with the urban cooperative banks in the recent past when 40 banks have been taken into liquidation without any disruption or crises. With extended mandate, the involvement of DICGC in recovery and problem bank resolution could help in reducing the cost to DICGC by having greater control on liquidation process and there by reducing moral hazard to a certain extent.

7.2.2 Ownership and governance

The overriding argument in favour of a government-run scheme is financial integrity. A privately run system may lack credibility without government support and enforcement mechanism, while a government backed system would be tempted to set low premiums for financial self-sufficiency under assurance that government will cover the financial gap. To price its product and to reduce moral hazard the private deposit insurer will have to have an information gathering arrangement with the bank whose depositors are being provided the cover.

The credibility of DICGC in India stems from the fact that it has a government backing as it has been created under an Act of Parliament and is wholly owned by the Reserve Bank of India. The operations of the DICGC are closely linked with the Reserve Banks regulatory functions for safeguarding interest of the depositors. The deposit insurance scheme basically aims at the protection of small depositors of banks, with the premium being paid by banks.

7.2.3 Hazard in deposit insurance

The deposit insurance scheme in India is an explicit scheme where the amount of cover is restricted to Rs 100000 per depositor, representing a cover of about 3.5 times of per capita GDP. Furthermore, only deposits with banks are covered. In case of commercial banks, they are subject to adequate regulation and supervision. In case of cooperative banks there have been problems arising out of dual regulation. In the wake of a shake out in this sector in 2002, the confidence in the urban cooperative banks was severely shaken. In the recent period, a clear cut policy of having agreements with the State Governments for strengthening the sector by having a coordinated and convergent approach for taking regulatory action has been brought in and this has slowly brought back confidence in the sector. The process of weeding out insolvent banks in a non disruptive fashion was greatly facilitated by the existence of the explicit paybox scheme of the DICGC with depositors of most of the UC Bs being covered by the DICGC. At the same time stiff regulatory action has been taken to avoid moral hazard problems.

7.2.4 Market discipline

Greater reliance on market discipline is a very important measure to address moral hazard. In India, Reserve Bank as regulator requires a high level of disclosure by supervised entities. Since 2005, penalties levied on banks are placed in the public domain. Most commercial banks are listed and the stock exchanges' regulations on disclosure further aid transparency. Significant deviations between audit reports and results of supervisory examination are taken very seriously and auditors are held accountable for disciplinary action. In the recent period, the Reserve Bank under the MOU with the State governments has insisted upon professional firms to audit the cooperative banks.

7.2.5 Insurance Premium and Deposit Insurance Funds

Ideally, the insurance premium applicable must be weighted by some objective measure of risk of each entity, so as to avoid a situation where entities at higher risk are subsidized by those operating with a lower degree of risk. In India, we have a uniform premium irrespective of the risk and there is a view, especially among the commercial banks that they are subsidizing the cooperative banks. Work is, however, under way to move over to risk based premium system.

There are two views on ex post and ex ante funding. It has sometimes been argued that the ex post financing of the scheme is preferable to the ex ante approach, because it is less onerous for the banks. On the other hand the interest foregone by banks on their contribution increases the value of the fund and feeds into lower premium. The Deposit Insurance Fund (DIF) of DICGC is funded by collection of premium from member-banks (ex-ante) and the fund is invested only in Central Government securities. The interest income from such investments and cash recovery of assets of failed banks also contribute to DIF. As on March 31, 2006 the Reserve Ratio stood at 0.86 pre cent. The DIF has, in the event of shortfall, a line of credit available from the Reserve Bank.

7.2.6 Reimbursing Depositors

A basic function of all deposit insurance systems is to reimburse depositors promptly for the losses they would otherwise suffer when a financial institution closes. Reimbursements to depositors including the speed and convenience of payment vary across countries, and can affect public confidence in the capability of the deposit insurance system. As per DICGC Act, the Corporation is required to settle the claim of the depositors within two months from the date of receipt of the claim list from the liquidator. The liquidator is required to submit the claim list of depositors within 3 months of taking charge as liquidator. However, it is observed that liquidators of cooperative banks generally fail to submit the claim within the statutory time frame of 3 months, resulting in delay in settlement of depositors claims.

Delays in settlement of claims of depositors are one of our major concerns. Delay in submission of claim list by the liquidators is mostly due to delay in audit of accounts, poor maintenance of records /non-availability of proper records, deficiency in the claim list and inclusion of ineligible deposit accounts resulting into protracted correspondence with the liquidators, non production of required information/documents to the Chartered Accountants deputed for verification of claim list , filing of appeals against the liquidation order, by the directors of the bank and other stake holders etc. The power to appoint the liquidator for the co-operative banks vests with Registrar of Cooperative Societies. It is observed that usually, one liquidator is appointed for more than one bank for liquidation work and this duty assigned to him is in addition to his other regular duties. Often the liquidators do not have expertise or knowledge about banking practices and working ng. They depend upon staff of failed banks and external agencies for preparation of claim list. In spite of detailed guidelines issued to the liquidators, the claim list is not prepared strictly as per the provision of DICGC Act 1961 and guidelines.

The Corporation has taken various steps to reduce the time lag for depositors to get their dues vz. (i) periodically reviewing the position of receipt, settlement and pending claims,

- appointment of CAs for verification and certification of claim lists prepared by the liquidators,
- reducing the time lag between the date of liquidation and date of submission of claims
- list of common deficiencies observed are being provided to the liquidators,
- conduct training programmes for the liquidators and State Government Auditors. Besides, the Corporation is also exploring ways for ensuring appointment of competent liquidators

7.3 Crop Insurance

7.3.1 Introduction

Risk and uncertainties are common in agriculture due to the very characteristic of agriculture i.e., dependence on nature. There are number of factors that affect the returns from farming, many of which are beyond the control of farmers. Occurrence of drought, flood, untimely rainfall, hailstorm etc. is only a few among the long list of factors that affect the returns from farming directly. The increase in the probability of occurrence of such extreme climatic events has worsened the status of farmers, who are at the receiving end. The occurrence of such untoward climatic incidents has become very normal, in recent years, as a consequence of global climate change.

India is considered to be vulnerable to the effects of climate change due to several factors like high dependence on agriculture, low coverage of irrigation, lower resource availability at individual farms and unavailability of proper technology to combat the risk. Other factors like dominance of small and medium sized holdings coupled with inherent lacunae like farmers' apathy towards newer technologies, unscientific post-harvest management, and un-organized and chaotic marketing system, also contribute to making agriculture a perilous endeavor in India.

So far, the country is not fully equipped to handle risk situations and, to insulate its farmers from both production and price risks. Farmers, on the other hand, are compelled to adopt adaptation strategies, both ex-ante and ex-post. Crop diversification, occupational diversification, staggered sowings, self-stabilization funds, contract farming etc. are some of the ex-ante strategies. Apart from these, farmers may also resort to the ex-post adaptation strategies like sale of assets, migration, loans (formal and non-formal), and consumption credits etc. The protection offered by these strategies, however, is not always sufficient since it depends on the nature and scale of risk incidence. Under such circumstances insuring crops is considered as a key strategy.

Crop insurance, which is an ex-ante adaptation strategy, can transfer the risk from the insured farmers to the insurer agency. By paying a nominal premium amount, it facilitates the farmers to receive larger compensations subject to the incidence of unforeseen loss. Insurance lends support to farmers after distress: compensation amount can help to start new agricultural season after a bad year. At the same time, it reduces the burden on Government's disaster payments. For an agrarian economy like India, crop insurance is an important part of agricultural policy.In crop insurance schemes, as in other insurance, the insurer agrees to pay the sum insured in the event of realization of underlying risk, on payment of prescribed premium. Insurance works on 'law of large number', i.e., risks are independently distributed rare events provided sample is sufficiently large. The insured will be paid compensation at the realization of risk by the premium collected from those who haven't faced the risk or from the accumulated premium of good years. Indeed, Insurance programme to be successful, risk diversification through pooling across space and time is essential.

7.3.2 Evolution

Considering the importance of publicly supported insurance scheme for agriculture, Government of India constituted the expert committee under the chairmanship of Dharam Narain to study the possibility of crop insurance schemes. The committee ruled out the possibility of implementing the scheme. In contrary, Prof. Dhandekar recommended starting insurance programme for farmers. Accepting the recommendations, Government of India set up General Insurance Corporation (GIC) in 1973.

GIC introduced a crop insurance scheme on pilot basis in Gujarat in the year 1973, which was later extended to West Bengal, Tamil Nadu and Andhra Pradesh. Later, in 1985, the first ever full-fledged crop insurance programme, the 'Comprehensive Crop Insurance Scheme' (CCIS), was introduced by GIC in most of the states. The scheme targeted to provide risk protection to all the loanee farmers. In this scheme, sum insured was limited to Rs. 10,000 per farmer irrespective of the magnitude of loan. The insurance premium was fixed at two per cent for cereals and one per cent for pulses and oilseeds. Despite these efforts, the scheme could not achieve its intended targets due to the flaws like limited crop coverage, capping on sum insured, and availability limited to loanee farmers.

The Government, then, decided to launch the National Agricultural Insurance Scheme (NAIS) in the year 1999. The scheme covered most of the crops, all the states and was available to both loanee as well as non-loanee farmers, with the difference that it is mandatory to the former. For an effective implementation of the scheme, Agriculture Insurance Company was also set up in the year 2002, with General Insurance Corporation of India (GIC) National Bank of Agriculture and Rural Development (NABARD) and four public-sector general insurance companies as its stakeholders.

Weather Based Crop Insurance Scheme (WBCIS) was then piloted across India from Kharif 2007, to explore the effectiveness of weather based crop insurance as an approach to overcome the constraints of National Agricultural Insurance Scheme (NAIS). WBCIS provides protection against losses in crop yield resulting from weather related risks. Weather parameters are used as an index for compensating crop loss based on the simulation models of crop yield and concerned weather parameters. Farmers get compensation if the insured weather parameters moves above or below the normal range. The scheme is bound with many limitations since it can be operated in only such areas where automatic weather stations are available, and can insure only parametric weather risks.

Subsequently, NAIS was modified and re-launched as Modified National Agricultural Insurance Scheme (MNAIS) and was implemented in selected states, on pilot basis. Unlike NAIS, MNAIS works in the regime of actuarial premium and has a broader definition of risk which includes cyclones and prevented sowings. In 2013, NAIS, WBCIS and Coconut Palm Insurance Schemes were merged into a comprehensive scheme- National Crop Insurance Programme (NCIP). This scheme is operation in all the districts from kharif, 2014.

7.3.3 Features

Crop insurance is one of the various types of insurance that are offered to the people. This insurance is directed to the farmers and agriculturists. This Crop insurance scheme has been going on since the time of Kharif 1985. This insurance offers financial assistance for risk management in agriculture.

This insurance policy is a relief scheme for the farmers whose crops get spoiled during natural catastrophe. The insurance amount that is offered to the farmers is equal to the loan amount that has been disbursed to them. A certain amount of premium is charged against the crop insurance.

The loss that is incurred due to natural calamities is met by the Government of India. It is to be noted that the insurance covers only one crop. The crop insurance does not cover financial assistance to multiple crops.

Experimental Crop Insurance Scheme is offered to the marginal farmers. This was introduced during the 1997 to 1998 rabi season by the Government of India. Large numbers of farmers are included under this scheme. Financial security is offered to the farmers by this crop insurance.

Crop insurance services are offered to the farmers for better production of the crops and introduction of modern technologies. Top quality services are rendered to the farmers and clients. The risk factor that is involved with the production of the crops has reduced much because of the introduction of this insurance policy.

7.3.4 Issues and challenges

Indian crop insurance works on 'area approach' i.e area is the basic unit for estimating yield loss and rate of compensation. The area approach works well only when correlation between farmers' yield and yield at crop cutting experiments are high. This condition is rarely met resulting in 'Basic Risk' i.e., farmer suffering crop loss but not receiving compensation. In addition, for the people who finally receive the compensation; the payment is delayed considerably in most of the cases. Loss sharing between center and state governments and operational formalities involved in it are sighted as the major reason for the delay.

Since the insurance market in agriculture sector is far from perfect, it needs to be bundled with other products like credit: which India has done. Whoever buys short term agricultural credit, they automatically and mandatorily insure their crops. But for a farmer, who wishes to take insurance product voluntarily, banks act as financial intermediaries. Experience till date suggest that the poor incentive that the banks have in promoting crop insurance is one of the reason for lesser coverage of insurance schemes. Only 4% of the premium is paid to bank as service charge, which is very low. This poor incentive along with the increase in work load makes the banks poor promoters of voluntary crop insurance.

Insurance is a financial tool to manage risk and not an investment to increase income. It also involves complicated terminologies like sum insured, indemnity levels etc. understanding of which is crucial for farmers to make informed choices. Awareness creation and training programmes play a crucial role in convincing farmers to take up insurance. Lack of adequate database for determining premiums and indemnities is another challenge that the insurance providers and policy makers face.

7.3.5 Opportunities

Even though hurdles exist in the way forward, intend should be to see the opportunities. The large pool of uninsured farmers existing and still continuing farming in our country is the biggest opportunity for any crop insurance scheme. In future, these millions of farmers will be compelled to insure at least some part of their crops, since the risk in farming may increase in future due to the effects of global climate change. Diverse climatic conditions across parts of India may be best of a bad bunch as it imparts some randomness to systematic risks (provided the insurance scheme is implemented throughout the country). The government have realized the level of distress in which the farmers are in, and this has led to a high policy thrust in favour of crop insurance. Finally, the good network of self-help groups and cooperatives existing in India can be used for micro-insurance, provided a good model of such kind is developed through proper planning and research.

7.4 Agricultural Credit

7.4.1 Introduction

The presence of informal lending as an alternative to formal credit is due to the adverse selection problem faced by lenders. Lenders have little to no way of distinguishing good borrowers from the bad-risk borrowers; as a result, good borrowers are rationed out of the formal market. Small farmers who may be good credit risks are unable to acquire credit from formal channels and resort to these informal sources. In the Philippines, these are comprised of the traditional moneylenders which were the landlords, as well as traders, input suppliers, big farmers and warehouse owners. Llanto recognizes that these informal lenders are able to resolve the problem of information asymmetry since the borrower and the informal lender know each other quite well such that the lender can be more accurate in assessing the likelihood of repayment and default. Also, the loan contract that is negotiated carries with it specific characteristics that account for a farmer's circumstances. These include flexible repayment schedules, varying amounts of loan amortization, and adjustment of the required collateral, among others.

A feature commonly associated with rural credit markets is the interlinking of transactions. This reduces the cost of acquiring information and reduces the transaction costs of the lender. Llanto cites such linkages between banks and rural organizations that are termed the self-help groups. These are associations organized, owned, operated, and controlled by the members themselves. Credit cooperatives, farmers associations, credit unions all fall under this category. The link or tie-up happens at an institutional level between the group and the bank. Llanto presents a case of a rural-based corporation in Cotabato which aids its farmer-members to acquire loans at market rates, acting both as a conduit of funds and a guarantor to the loans. The corporation is able to reduce transaction costs since it is able to identify the good risk borrowers among its members. In cases of default, the corporation is able to exact pressure on members to pay back their loans to ensure that credit access to members will be uncut.

7.4.2 Federal Outlook

The federal government has a long record of supporting agricultural credit markets, dating back to 1916, when Congress chartered the Federal Land Banks, the first component of the Farm Credit System (FCS). Today, three institutions have primary responsibility to deliver federal farm credit policies. The USDA's Farm Service Agency (FSA) provides direct loans and loan guarantees to farmers unable to meet conventional credit standards. The cooperative FCS is a government-sponsored enterprise (GSE) with an implicit federal guarantee, which serves farmers that meet conventional lending standards.

The privately owned Federal Agricultural Mortgage Corporation (Farmer Mac), also a GSE, provides liquidity for rural lenders by buying agricultural and Steven R. Koenig, Economic Research Service, USDA Damona G. Doye, Oklahoma State University Agricultural Credit Policy rural home loans. Other federal lenders with less prominent roles serving agricultural producers include the Small Business Administration, the Commodity Credit Corporation, and the Rural Business and Cooperative Service. The Federal Home Loan Bank System (FHLBS) also serves as a source of funds for rural lenders. Roughly one-third of the $180 billion in farm debt at the end of 2000 was held by or guaranteed by the government, through FSA or GSEs. In some regions or for some classes of borrowers, the federal government supplies or supports half of all farm credit needs.

Federal credit subsidies may have different objectives, but they all work to lower production costs for farmers and increase access to credit. Today, direct and indirect federal farm credit subsidies are substantial — totaling hundreds of millions of dollars annually. In addition to specific farm credit policies, the regulation and supervision of financial institutions, as well as monetary, fiscal, tax, and antitrust policies have a significant effect on the operation of agricultural credit markets.

The credit titles in the last three farm bills have dealt mostly with FSA farm loan policies, and their content reflects the issues of those periods. The 1985 Farm Bill was drafted during a farm economy downturn and emphasized help to farm borrowers unable to repay their loans or assist after failure.

Credit Policy Issues and Options

Federal farm credit policies over the past 85 years have been motivated by perceptions that private lenders were not adequately, efficiently, or fairly supplying credit to farmers. Federal credit policies seek to address concerns about social equity and to enhance economic efficiency by raising lender competition, lowering transaction costs, or improving market information. Some specific policy objectives have included raising farm-family incomes, overcoming the difficulties of financing farm startups, stemming the out-migration from farming, reducing the financial impact of natural disasters, decreasing interregional variation in borrowing costs, improving liquidity to rural capital markets, and reducing credit rationing.

Today, Congress is faced with defining a proper role for federal farm credit subsidies as structural change in agricultural production and within the financial services industry continues at a rapid pace. Past federal farm credit programs and policies may be inappropriate or ill equipped to address issues relevant to a modern and increasingly industrialized agriculture. In this context, policy makers may address the role of federal credit policies in providing a farm financial safety net, affecting farm financial market performance, and in addressing concerns of social equity and structural change. These and any other policy objectives will likely have to be balanced against concerns over federal credit subsidy costs.

Credit facilitates the acquisition of production inputs. In the context of the World Trade Organization, any new proposals calling for additional government farm credit subsidies would likely be considered trade-distorting domestic support. As such, some policies and proposals would likely be in the "amber box" domestic support category.

Providing a Farm Financial Safety Net

USDA subsidized farm loans have traditionally been used to aid financially stressed farmers, especially during economic downturns or in response to losses incurred from natural disasters. Weak commodity prices have significantly lowered farm revenues in recent years. To maintain aggregate farm income at a politically acceptable level, the federal government is making record amounts of payments to farmers. These payments have lessened the need for credit and enhanced farmer creditworthiness.

While most federal farm aid is transferred through other means, farm loan subsidies delivered through the Farm Service Agency remain a significant component of the federal farm financial safety net. Congress more than doubled FSA's farm lending authority from fiscal 1998 to fiscal 2000 to $5.6 billion to assist the least creditworthy producers.

Federal farm loan programs are discretionary expenditures and, as such, are subject to annual federal budget allocations. While recent U.S. budget surpluses have reduced the urgency to cut federal outlays, the desire to trim costs and streamline loan programs remains an important policy objective.

Farm Service Agency Farm Loan Programs

Congress could elect to expand the funding or eligibility of FSA loan programs to assist a broader range of farmers. This policy would raise subsidy costs and, in the absence of sufficient needs testing or targeting, would likely be more of an income transfer mechanism as opposed to a financial assistance mechanism. If income transfer is the objective, other mechanisms can be more effective. Alternatively, Congress could elect to cut farm loan program funding and/or their scope of eligibility. Limiting funding or eligibility would mean that fewer borrowers would qualify for federal credit and, hence, would face greater credit costs or an inability to obtain credit.

Coping with Structural Change and Helping Under-served Groups

Structural change in U.S. agriculture has hastened in recent years, especially for certain enterprises. Federal credit subsidies influence structural change by reducing the cost of capital and funneling subsidized capital to specific groups. The industrialization of agriculture is producing a dualistic structure where a relatively small number of large farms produce the majority of food and fiber, and a large number of small farms produce very little. Midsized farm numbers are falling and this raises concerns that farming is evolving in directions that may lead to the demise of the "family farm."

Preserving the family farm remains at least rhetorically a policy objective. The average age of farmers rose one full year to 54 years from 1992 to 1997, while the number of young farmers continued to fall. These trends are raising concerns that an insufficient number of new farmers will be available to replace retiring farmers. The large amounts of capital required to operate a cost-competitive farm can be a significant barrier to entry for new farmers.

Farm Service Agency Targeting

FSA farm loans currently target family-sized farmers, beginning farmers, and under-served borrowers to help them obtain an economically sufficient size. Targeting is largely accomplished through caps on the amount of FSA credit a borrower can obtain, or by allotting a share of annual lending authorities for targeted groups. Better targeting methods could be employed and more incentives provided to private lenders to increase guaranteed loan program targeting.

However, while providing more credit subsidies reduces costs for qualified applicants, this by itself may be insufficient to have a significant effect on the policy objective being pursued, as many other factors may be more important. Evidence also suggests that non-targeted groups, frequently through higher bid prices for business assets, may capture a portion of the subsidy provided.

Enhancing Agricultural Credit Market Performance

New technologies, regulatory reforms, and other factors are reshaping the financial services industry. These changes are reducing economic barriers, increasing market integration, enhancing liquidity, and, therefore, minimizing the need for government intervention in rural credit markets. Despite structural change within the financial services industry and within agriculture itself, concerns linger that some rural financial markets might still face a shortage of reasonably priced loan funds. For example, bank consolidations and mergers raise fears that the commitment of local institutions to agriculture will be lessened.

USDA's report, titled Credit in Rural America, concluded that rural financial markets work reasonably well in serving the financial needs of most sectors of the rural economy. The report went on to state "policies that provide untargeted subsidies to a broad range of rural lenders or borrowers, such as examined in this report, are unlikely to be cost effective." Nonetheless, legislative and regulatory policy changes continue to address concerns about credit market liquidity, competitiveness, and efficiency.

7.5 Recent development in agricultural credit

7.5.1 Introduction

Credit supply is an important determinant of investment in agriculture. Since the nationalisation of commercial banks in 1969, India had strongly pursued a policy of "social and development banking" in the rural areas. As a result, formal institutions of credit provision, mainly commercial banks, emerged as important sources of finance to agriculture displacing usurious moneylenders and landlords. The policy of social and development banking was a supply-led policy; it aimed at augmenting the supply of credit to rural areas, and that too at an affordable interest rate.

Three aspects of the post-1969 policy of social and development banking stand out:

- First, according to the new branch licensing policy, commercial banks were required to open four branches in unbanked rural areas for every branch opened in metropolitan or port areas. As a result, if there were only 1443 rural branches of banks in 1969, there were 35,134 rural branches of banks by 1991.
- Secondly, according to the policy of priority sector lending, 40 per cent of the net bank credit was to be compulsorily provided to those sectors of the economy (or sections of the society) that would not get timely and adequate credit in the absence of binding targets. These sectors were, typically, loans to farmers for agriculture and allied activities (18 per cent), micro and small enterprises, poor people for housing, students for education and other low income groups and weaker sections (10 per cent).
- Thirdly, according to the differential interest rate scheme of 1974, loans were provided at concessional interest rates on advances made by public banks to selected low income groups to engage in productive and gainful activities. The differential rate of interest was fixed uniformly at 4 per cent per annum, i.e. 2 per cent below the bank rate.

There is little quarrel among economists on the effect that the increased flow of credit after 1969 had on agricultural growth in India. Increased availability of credit from public banks helped small and marginal farmers adopt costlier new technologies and farming practices, which were a part of the green revolution strategy.

Yet, in the early-1990s, the policy of social and development banking was criticised by the proponents of financial liberalisation. Each of the three aspects of rural credit expansion listed above were sought to be undone. The Committee on the Financial System (Narasimham Committee) argued that banks should function on a commercial basis, and profitability should be the prime concern in their activities. Thus, banks were to be permitted to close rural branches, in the name of rationalisation of branch networks. Priority sector norms were increasingly diluted. It was argued that banks should be given a free hand to charge rates of interest as administering interest rates would lead to financial repression.

As a consequence, the period of financial liberalisation after 1991 was a period of reversal of the achievements of social and development banking. It is by now well documented that the trends that emerged in India in the 1990s with respect to the supply of rural credit in general, and agricultural credit in particular, were deeply disturbing. In the 1990s, there was

- large scale closure of commercial bank branches in rural areas;
- a widening of inter-State inequalities in credit provision, and a fall in the proportion of bank credit directed towards regions where banking was historically underdeveloped;
- a sharp fall in the growth of credit flow to agriculture;
- increased sidelining of small and marginal farmers in the supply of agricultural credit;
- increased exclusion of the disadvantaged and dispossessed sections of the population from the formal financial system
- strengthening of the hold of moneylenders on rural debt portfolios.

In 2004, the government announced its intent to double the flow of credit to agriculture over a period of three years. Increase in credit flow was an integral part of the so-called "New Deal for Rural India" promised by the United Progressive Alliance (UPA) government. A comprehensive credit policy was announced in June 2004, which included the commitment to raise agricultural credit flow by 30 per cent every year, financing of 100 farmers per branch (thus, 50 lakh farmers in a year), two to three new investments in agricultural projects per branch every year and a host of debt relief measures, such as debt restructuring, one-time settlement and financial assistance to redeem loans from moneylenders. From 2004 onwards, it is regularly claimed in official circles that the flow of credit to agriculture has been increasing at a rapid rate, even surpassing its annual targets. In fact, an impression is often gained from official statements that the problem of agricultural credit has been set right with the doubling of credit flow, and the concurrent expansion of micro-credit.

7.5.2 Trends

Historically, agricultural credit has comprised mainly of credit provided directly to cultivators, which was called "direct finance to agriculture". Within direct finance to agriculture, short-term- credit or credit for seasonal agricultural operations has accounted for a significant share. Short-term loans to agriculture are referred to as "crop loans", as they are advanced for crop cultivation against the hypothecation of the crop to be cultivated by the farmer. Crop loans are provided as cash or in kind, such as the supply of fertilisers and seeds. Apart from crop loans, direct finance also includes credit for medium and long-term investment in agriculture. The second component of agricultural finance is called "indirect finance", which does not go directly to cultivators but to institutions that support agricultural production in rural areas. The typical forms of indirect finance to agriculture were loans to input dealers for their role in the provision of agricultural inputs and loans to electricity boards for supplying power to cultivators.

7.5.3 Urbanisation of agricultural credit

There was an increased provision of agricultural credit from bank branches in urban areas in the 2000s. The share of agricultural credit in rural areas, which are commonly associated with agricultural activities, saw a fall during this period. In 2011, about one-third of total agricultural credit and one-fourth of direct agricultural credit was outstanding in bank branches located in the urban areas. This again implied a relative diversion of agricultural credit towards urban-based dealers (as part of indirect credit) and urban-based corporate (as part of direct credit) and away from farmers based in rural areas.

7.5.4 Increasing disconnect between credit and investment

The changes in the nature of agricultural credit in the 2000s also led to a growing disconnect between credit and investment in agriculture. First, there was a sharp fall in the share of long term credit (that aids investment) and a sharp rise in the share of short term credit in the lending of commercial banks of India after 1991. In 1990-91, the share of long term credit in total agricultural credit was about 66 per cent, which fell to 44.4 per cent in 2008-09. On the other hand, the share of short term credit in total agricultural rose from 34.1 per cent in 1990-91 to 55.6 per cent in 2008-09. The sharp decline in lending of long term loans is likely to have significantly contributed to the fall of agricultural investment in the 1990s and 2000s.

Credit and capital formation were roughly at similar levels, as well as they moved together at every year. However, after 2001-02, trends in credit and capital formation began to diverge. From 2002-03 onwards, agricultural credit grew rapidly compared to capital formation in agriculture, and the difference between the amounts of agricultural credit and agricultural capital formation grew. In other words, only an increasingly smaller portion of credit supplied to agriculture was transformed into capital investment in agriculture in the 2000s.

8

Preparation and Analysis of Financial Statements

8.1 Financial Statement Analysis

A financial analysis of the farm operation is done to determine the financial position and performance of the business. The financial analysis of an agricultural business must focus on both its present position (called its financial position), the results of its operations, and its past financial decisions (called financial performance). The farm's financial position refers to the total resources controlled by the farm business and the total claims against those resources at a single point in time. Measures of the financial position provide an indication of the capacity of the farm business to withstand risks. It also provides a benchmark to compare with the results of future farm business decisions. The financial performance refers to the results of the farm business's production and financial decisions over one or more periods of time. Measures of financial performance include the impact of external forces that are beyond anyone's control (i.e., drought, state health regulations that restrict the sale of goat milk or foreign imports of frozen goat meat) and the results of the farm's operating and financial decisions made in the ordinary course of business.

Financial Statement Preparations

- ❖ Prepare Income Statement: The income statement calculates the net profit or loss, which are determined by sales revenue, expenses, and general ledger records.
- ❖ Prepare Statement of Retained Earnings: The statement of retained earnings reflects the distribution of profit between retained earnings and dividends.
- ❖ Prepare Balance Sheet: The balance sheet illustrates the company's financial position – it's a snapshot of the last day of the accounting cycle.

- Prepare Cash Flow Statement: From a cash basis, the statement of cash flow is a compilation and comparison of information from the three main financial reports: income, retained earnings, and balance sheet.
- Financial Statement Analysis: Business owners may not need all the details, but they definitely need to understand the big picture of their current financial and cash positions. These insights are crucial when making business operation decisions.
- File Financial Statement Reports: All companies, foreign and domestic, are required to file registration statements, periodic reports, and other forms electronically through the Electronic Data Gathering, Analysis, and Retrieval system. The purpose of EDGAR is to increase the efficiency and fairness of the securities market for the benefit of investors, corporations, and the economy via a time-sensitive collection tool.

8.1.1 Analyzing operations

- Determine the objectives of the analysis . Is the analysis being conducted to determine the tax liability? Is the analysis being done to apply for credit? Is the analysis being done to determine the health/profitability of the operation?
- Describe the business organization and its goals . Is the operation expected to make a profit? Are there measurable goals with regard to profit and growth? Who is in charge of analyzing the business?
- Prepare financial statements. Be sure to exercise be consistent between years so the comparisons will be valid. Make sure that the data is accurate and complete. Be sure to include accrual adjustments.
- Calculate the financial ratios and prepare historical and projected financial summaries. Again, check for consistency and accuracy.
- Compare your farm's results to similar operations if benchmarks are available .
- Summarize the analysis to help when reviewing at a later date . Strengths and weakness should be expressed in a clear, concise manner in a way that the reader can use and see any limitations of the analysis.

8.1.2 Financial Measures and Ratios

The two balance sheet measures most often used to evaluate Liquidity are the current ratio and working capital. The current ratio is the relationship between current farm assets and current farm liabilities. It is calculated as:

Current Ratio=Total current assets ÷ Total current farm liabilities

The ratio indicates the extent to which current farm assets, if liquidated, would cover current farm liabilities. The higher the ratio, the greater the liquidity. If the ratio is greater than 1.0, the operation is considered liquid. If the ratio is less than 1.0, the operation has some degree of cash flow risk. Generally, lenders and financial analysts like to see a current ratio of 1.5 to 2.0. One consideration when calculating the current ratio is deferred taxes. Because the ratio determines the impact of selling all current assets, the tax consequence should be considered. It is therefore a more conservative approach to include deferred taxes as a current liability when calculating the ratio.

Working capital is a measure of the amount of funds available after the sale of all current assets and the payment of all current liabilities at a single point in time. It is calculated as:

Working Capital = Current farm assets – Current farm liabilities

Because working capital is expressed as a dollar amount, it is difficult to make comparisons between operations. Generally, working capital should be positive, but the amount needed depends on the type and size of the operation, the time of the year, and the related seasonality of the production cycle.

Measures for Solvency

Solvency indicates the farm business's ability to repay all of its debts if all the assets were sold. If the value of the total farm assets exceeds the value of the total farm liabilities, the farm is said to be solvent. If the sale of all the farm assets would not be enough cash to pay off all liabilities, the farm is insolvent. The difference between the value of total farm assets and total farm liabilities, referred to as net worth or owner's equity, is the most often used measure of solvency. The most realistic approach to calculating solvency (owner's equity) is to use the market-based value of the assets, including the consideration of deferred taxes. There are three commonly used ratios to measure financial solvency are the equity-to-asset ratio, the debt-to-asset ratio, and the debt-to-equity ratio. All three of these ratios are related and neither is necessarily preferred. The equity-to-asset ratio measures the proportion of total farm assets owned or financed by the owner's equity capital. It is calculated as:

Equity-to-Asset Ratio = Total farm equity ÷ Total farm assets

The higher the equity-to-asset ratio, the more capital is supplied by the farm owner and the less is supplied by creditors. There is no exact standard for the equity-to-asset ratio that should be applied to every farm business. However, as the percent equity increases above 0.50, the owner is supplying a greater percent of the total assets in the business than the creditors. This ratio should increase over time if the owner retains farm profits and reduces debt obligations.

The debt-to-asset ratio measures the proportion of total farm assets owed to creditors. It is calculated as:

Debt-to-Asset Ratio = Total farm liabilities ÷ Total farm assets

The higher the ratio, the greater the risk to the farm business and those who are providing loan funds. The operator has less flexibility to respond to adverse natural or market conditions. As with the equity-to-asset ratio, there is no exact standard for every farm business. However, a debt-to-asset ratio greater than 0.50, indicates that the owners contribute less than 50 percent of the value of the farm's assets. In with this situation, creditors are often cautious about making loans.

The debt-to-equity ratio measures the proportion of funds invested by the creditors versus the farm owners. It is calculated as follows:

Debt-to-Equity Ratio = Total farm liabilities ÷ Total farm equity

The higher the debt-to-equity ratio, the more total capital is supplied by the creditors and less by the farm owner. This ratio is also referred to as the leverage ratio. Leveraging refers to increasing the use of debt relative to equity as a means of financing the business. Lenders are particularly interested in this ratio because it shows the proportion of the risk they are taking in comparison to the owner. Many lenders prefer the debt-to-equity ratio to be less than 1.0, with requirements varying depending on whether the liabilities are secured by current, intermediate, or long term assets. In general, the greater the loan's risk and the longer the loan's term, the lower the lender wants the debt-to-equity ratio to be.

Measures for Profitability

Profitability measures the financial performance of the operation over a period of time, usually one year, that result from the decisions made regarding the use of land, labor, capital and other management resources. The five commonly used measures to assess profitability are net farm income, net farm income from operations, rate of return on assets, rate of return on equity, and the operating profit margin. The rate of return on assets (ROA) measures the relative income generated by the assets of the farm business and is often used as an overall index of profitability. The ROA is calculated as follows:

ROA = (Net farm income from operations) + (Farm interest expense) – (Value of unpaid operator and family labor and management)/Average total farm assets

Once the income statement has been developed, the net farm income from operations and the farm's interest expenses can be taken directly from the income statement. The value of unpaid operator and family labor and management must be estimated. Withdrawals from the business for family living expenses can be used to estimate unpaid operator and family labor and management. The average total farm assets can be calculated by adding total assets from the beginning balance sheet plus total assets from the ending balance sheet and dividing by 2. This ratio is often used as an overall index of profitability. It is best to use the cost basis approach when evaluating your individual farm business over time because market based values fluctuate over time. But, when comparing your farm to other farms, it is best to use the market value approach to value the farm assets, because cost basis values can cause extreme differences between farm businesses.

The rate of return on assets will vary among different types of agricultural operations, but the higher the ROA, the more profitable the operation. While the ROA is most often compared across years within an operation, the ROA for any particular year can also be compared to the average interest rate the operation is currently paying or to the cost of new borrowing. If the ROA exceeds the interest rate of the new proposed borrowing, then borrowing more can be used to profitably grow the business (equity). However, if the ROA is less than the average interest rate that the operation is currently paying, then borrowed funds are not being used profitably, and adding new debt will reduce the growth of equity. Therefore, the level of profitability is an important key to the successful use of debt financing as a strategy to increase the equity of the operation. It should be noted that the ROA in most agricultural operations might seem low when compared to non-agricultural investments such as stocks and bonds. This is important and re-enforces the notion that people invest in agricultural operations for reasons other than profit and equity growth.

The rate of return on equity (ROE) is another measure used in determining financial performance or profitability. The ROE is calculated as follows:

ROA = (Net farm income from operations) – (Value of unpaid operator and family labor and management)/Average total farm equity

As with the previous calculation, the net farm income from operations and farm interest expense can be taken directly from the income statement, while the value of unpaid operator and family labor and management must be estimated. The average total farm equity can be calculated by adding the total farm equity from the beginning balance sheet plus the total farm equity from the ending balance sheet and dividing by 2.

Measures of farm Efficiency

There are a number of ratios that measure efficiency, which is an important component of profitability. The ratios relate physical output to selected physical inputs and help evaluate whether or not the farm assets are being used efficiently to generate income. The efficiency measures most widely used in agricultural businesses are the asset turnover ratio and the four operating ratios: the operating expense ratio, the depreciation expense ratio, the farm interest expense ratio, and the net farm income from operations ratio.

The Asset Turnover Ratio measures how efficiently farm assets are being used to generate gross revenue. Consideration should be given to the way in which the assets are valued and the same approach used to calculate the ROA should be used to calculate the asset turnover ratio. It is calculated as follows:

Asset Turnover Ratio = Gross farm revenue/Average total farm assets

The higher the ratio, the more efficiently assets are being used to generate revenue. The agricultural industry as a whole tends to have both a slow rate of asset turnover and a relatively low operating profit margin.

The Operating Expense Ratio reflects the extent to which gross farm revenues are expended on farm operating inputs, excluding depreciation and interest. The higher the value of the ratio, the larger the proportion of gross farm revenues is needed to offset all of the operating expenses. Ratios in the 40 to 60 percent range indicate relative efficiency, with efficiency declining as the ratio rises. The operating expense ratio is also used as one of four operational ratios.

Operating Expense Ratio = (Total operating expenses) – (Depreciation expenses)/Gross farm revenues

The Depreciation Expense Ratio measures the proportion of gross farm revenues that are represented by the depreciation expense (a non-cash expense). A relatively low depreciation expense ratio could indicate little difficulty in making planned and timely replacements of capital assets, or it may indicate that capital assets (usually farm machinery) are relatively old. It should be noted that IRS depreciation rules could distort this ratio and you should use management depreciation. This is also used as one of four operational ratios.

Depreciation Expense Ratio = Depreciation expense/Gross farm revenues

The Interest Expense Ratio measures the proportion of gross farm revenues that are required to cover the farm's interest expenses. Large interest expense ratios are characteristic of highly leveraged operations. As a general rule, the interest expense ratio should be less than 0.15. Interest expense ratios over 0.15 indicate that farm's interest expenses are a large proportion of its gross revenues and that the farm is likely suffering "financial stress." The farm interest expense ratio has important implications for the profitable use of debt financing and financial risk. As indicated in earlier discussions of profitability, if the rate of return on farm assets (ROA) exceeds the cost of debt financing, increasing debt can increase the growth in farm equity. The interest expense ratio is also used as one of four operational ratios.

Interest Expense Ratio = Total farm interest expense/Gross farm revenues

The Net Farm Income from Operations Ratio measures the net farm income as a proportion of its gross revenues. Thus, it reflects the proportions of gross farm revenues that remain after the farm operating expenses have been paid. It is calculated on a before-tax basis. The net farm income from operations ratio is also used as one of four operational ratios.

Net Farm Income from Operations Ratio = Net farm income from operations/ Gross farm revenues

The four operational ratios discussed above (b. through e.), when added together, should equal to 1.0 or 100 percent. The producer should always keep in mind that all of these ratios can vary widely between different operations and from year to year within an operation due to different types of farms and different marketing and production systems. Therefore, it is important that farmers compare projected values for the coming year to the most recent averages for their own operation.

8.2 SWOT analysis

SWOT analysis (Strengths, Weaknesses, Opportunities, and Threats) is a process that can help get insights into your business. For a SWOT analysis to work well, every member of your team (your family and/or your employees, your lawyer, your accountant, and your insurance agent) must be involved in the process.

8.2.1 Strengths

Strengths are considered mostly internal. For example, what do you and your family, employees, and management team bring to the business. If you are planning to start a business, you may have previous experience in the industry you are considering. If not, this may be viewed as a weakness. You may have an employee or member of your management team who has prior experience in the industry you are planning to enter. You may already know where to go to find the assistance you need and this would be considered a strength for this analysis.

If you are an existing business, you may have many years of expertise in the business. You know the history of your business and how you achieved the success you have. Your family is very supportive of the time you spend in the business and you have hired or developed experienced, loyal, and caring employees. Your management team works well together and provides excellent input for the business. Your equipment is up-to-date and you have saved the depreciation money to replace outdated equipment.

8.2.2 Weaknesses

Weaknesses are also considered mostly internal and are the items you will need to address for success of your business. If you are a start-up business, an example may be a lack of experience in the selected industry. Another consideration may be that the family members do not completely support you in this venture. You may have a lack of qualified employees or your business will not support full time employees, these would also be considered weaknesses. If you identify these as weaknesses, do not fear because existing businesses may have similar weaknesses.

Your current business may identify additional weaknesses such as, a previous poor decision that is now impacting the business. You may have a high turnover of employees, a lack of necessary capital for equipment, and/or a high short term debt load. These are all examples of weaknesses in an existing business. These may be the result of management decisions or influence from outside of the business, whatever the reason they will need to be addressed so your business can be successful.

8.2.3 Opportunities

Opportunities are considered mostly external. What opportunities are available for your business? If you are a potential business, you may take advantage of low interest loan packages from various sources for start-ups or possibly grants for feasibility studies. You may have an idea that has not been previously tried in your area and you know people are seeking the product you plan to produce. Potential businesses tend to be much more optimistic than existing operations.

Existing businesses have the decided benefit of viewing the industry from experience. You should know that if certain things happen in the industry, the results are usually predictable. For example, if you plant your crops before a certain date, germination or harvest will normally occur within a specific time frame. If for example, your area experiences a drought during the summer months, you may need to harvest the corn crop for silage before the crop enters a specific growth stage. You may also know where to sell corn silage from previous experience with droughts. You are usually better prepared to take advantage of opportunities more quickly due to your experience and may be better capitalized and able to withstand set backs.

8.2.4 Threats

Threats are also considered mostly external. These are threats from outside of your business that will directly impact you over which you may have very little control. If you are starting a new business, you may have the threat of local regulations negatively impacting your business. Unforeseen competition (local or foreign), dissolution of markets, and adverse weather may also have a negative effect on your new business.

Many of the previously mentioned threats will also negatively impact an existing business. Rising interest rates may have a greater impact on existing businesses than start-ups. Heavily leveraged businesses will be greatly impacted by a one or two percent increase in your interest rates.

When conducting your SWOT analysis keep in mind that one item can offset another. For example, your strengths may combat your weaknesses and your opportunities can offset your threats. If this does not happen, this exercise will enable you to point out these dilemmas so you may address them. The following diagram will serve as a guide for your thinking during the process.

8.2.5 Human Resources

- Hiring employees that have the skills you deem necessary is a way for you to combat a weakness. For example, if you need someone to drive a large truck, hiring someone with a commercial driver's license will eliminate that weakness. Another way to alleviate this same weakness is to learn how to drive the truck yourself. However, do you have the time to be away from the business for the time required to do the driving?
- Adding to your management team is another way to eliminate a perceived weakness. If your business is growing and you are handling the bookkeeping tasks yourself and this is becoming too much for you to do, hiring a bookkeeper will allow you to better focus on other parts of the business. If you do this, keep in mind to hire someone you trust or do an extensive background check, especially if they will have check signing capabilities. Having that person bonded will also create a level of safety.

8.2.6 Insurance

Insurance is another method of dealing with many of the issues you face. For many agricultural businesses, insurance is a method of eliminating perceived weaknesses, threats, or risks. If your analysis shows that you are under capitalized to handle a major emergency (as most agricultural businesses are), insurance coverage will help eliminate this issue. Threats from liability, fire, automobile accidents, and crop failures can all be reduced with insurance coverage.

Your liability insurance will cover anyone entering your farm and becoming injured. This coverage is included on your farm owner's policy but, the coverage is only for the dollar amount listed on the policy. For example, if your liability limit is $500,000 and someone sues you for $1M, you will be covered for the $500,000. The remainder of the judgment award is your responsibility. You should review your risk level and contact your insurance provider.

Fire insurance is also included on your farm owner's policy. You may choose to insure some buildings but not all. The level at which your policy will reimburse you is determined and is written in your policy. There are several methods of coverage. These consist of actual cash value, replacement value, and functional replacement value. You can find more explanation about this in the article Agricultural Business Insurance .

Your automobile insurance will be a separate policy from your farm owner's insurance. Any licensed over-the-road vehicle you own should be insured. If you have enough vehicles on the road, you may be eligible for fleet insurance which may save you some expense. Check with your Department of Transportation and your insurance provider to determine the level and coverage that is right for you.

Crop insurance coverage is available in many forms. You may cover your crops with several types of policies such as, Multi Peril (MPCI) and Crop Revenue Coverage (CRC), among others. You may insure your revenue with Adjusted Gross Revenue coverage (AGR) or Adjusted Gross Revenue-Lite coverage (AGR-Lite). Contact your crop insurance provider to determine the best coverage for your operation. More information regarding crop insurance can be found at Insert agent locator or by contacting tour local crop insurance provider.

9

Agricultural Cooperation

9.1 Meaning

Cooperatives are present in all the countries and in almost all the sectors, including agriculture, food, finance, health care, marketing, insurance & credit. A cooperative is an autonomous association of persons united voluntarily to meet their common economic, social and cultural needs and aspirations through a jointly owned and democratically controlled enterprise. Agriculture continues to be the engine of economic growth in most developing countries. Strong cooperatives are able to overcome many of the difficulties faced by the farmers; wherein in a country like India 85% are small and marginal farmers. Cooperatives work on the set seven principles, among these principles the seventh principle - Concern for the community derives cooperatives to work for the sustainable development of the community. There are a number of successful cooperatives in India itself like IFFCO (Indian Farmers Fertiliser Cooperative Ltd.) and KRIBHCO (Krishak Bharti Fertiliser Cooperative Ltd.) in the fertiliser sector, the Institution like AMUL in Dairy Sector and Self Help Groups (SHGs) formed by various Institutions have immensely benefitted farmer members in increasing crop productivity and overall income by generating support in various programmes related with agriculture. Suitable Farming Systems to generate year round employment and sustainable income through crops, vegetables, fruits and livestock have been achieved by the members of the cooperatives. Initiative has been taken to go in for Agro Forestry combining plantation of fruit trees, fuel trees, and forest trees to improve the overall climate on wastelands by cooperatives like IFFDC (Indian Farm Forestry Development Cooperative Ltd.). This has helped ecological resilience and rural livelihood enhancement among the community. The farmer members have been immensely benefitted in increasing efficiency of various agri inputs and overall crop productivity and finally making better profit through the efforts of cooperatives.

9.1.1 Principles

The Principles of Cooperation are guidelines by which cooperatives put their values into practice.

- Voluntary and Open Membership: Cooperatives are voluntary organisations, open to all persons able to use their services and willing to accept the responsibilities of membership, without gender, social, racial, political or religious discrimination.
- Democratic Member Control: Cooperatives are democratic organisations controlled by their members, who actively participate in setting their policies and making decisions. Men and women, serving as elected representatives, are accountable to the membership. In primary cooperatives, members have equal voting rights [one member one vote], and cooperatives at other levels are also organised in a democratic manner.
- Member Economic Participation: Members contribute equitably to, and democratically control, the capital of their cooperative. At least part of that capital is usually the common property of the cooperative. Members usually receive limited compensation, if any, on capital subscribed as a condition of membership. Members allocate surpluses for any or all of the following purposes: developing their cooperative, possibly by setting up reserves, part of which at least would be indivisible; benefiting members in proportion to their transactions with the cooperative; and supporting other activities approved by the membership.
- Autonomy and Independence: Cooperatives are autonomous, self-help organisations controlled by their members. If they enter into agreements with other organisations, including governments, or raise capital from external sources, they do so on terms that ensure democratic control by their members and maintain their cooperative autonomy.
- Education, Training and Information: Cooperative provide education and training for their members, elected representatives, managers, and employees, so that they can contribute effectively to the development of their cooperatives. They inform the general public – particularly young people and opinion leaders – about the nature and benefits of cooperation.
- Cooperation among Cooperatives: Cooperatives serve their members most effectively and strengthen the Cooperative Movement by working together through local, national, regional and international structures.
- Concern for Community: Cooperatives work for the sustainable development of their communities through policies approved by their members. It undertakes need based community-related services e.g., drinking water, health care, sanitation, education, women empowerment, and other community related programmes. A cooperative is a central rallying point in

the village and therefore it performs a lot towards the social responsibility and such activities bring the general community closer to the cooperative.

9.2 Agriculture cooperatives in India

Agriculture Co-operatives organise, promote & market, processing and storage of agricultural, horticultural and forest produce. They distribute agricultural machinery; undertake wholesale, retail and foreign trade. Effectively, major agriculture processing, distribution, supplies and agro-marketing across India are carried on through co-operatives.

1. National Co-Operative Development Corporation

It was established by an Act of Parliament in 1963 as a statutory corporation under the Ministry of Agriculture.

Functions:

Planning, promoting and financing programmes for production, processing, marketing, storage, export and Import of agricultural produce: and minor forest produce through co-operatives, besides income generating stream of activities such as poultry, dairy, fishery semi-culture, handloom etc.

NCDC Act has been further amended. It will now be able to finance projects in the rural industrial co-operative sectors and for certain notified services in rural areas like water conservation, Irrigation, and micro irrigation, agri-insurance, agro-credit, rural-sanitation, animal health etc.

Loans and grants are advanced to State Governments for financing primary and secondary level co-operative societies and direct to the national level and other societies having objects extending beyond one state. Now, the corporation can also go in for direct funding of projects under its various schemes of assistance on fulfillment of stipulated conditions.

2. National Agricultural Co-operative Marketing Federation of India

The objects of the NAFED shall be:

- to organise, promote and develop marketing processing and storage of agricultural, horticultural and forest produce.
- distribution of agricultural machinery, implements and other inputs, undertake inter-state, import and export trade, wholesale or retail as the case may be.
- to act and assist for technical advice in agricultural production for the promotion and the working of its members and co-operative marketing, processing and supply societies in India.

3. Tribal Co-Operative Marketing Development Federation of India Ltd

Objective:

- ❖ To secure higher earnings and generate employment opportunities for the tribal people of the country.
- ❖ To create awareness of the interplay of market forces among tribal.
- ❖ To provide assured markets and remunerative prices for tribal produce and also to undertake price support operations wherever required.
- ❖ To provide marketing and financial support to state level tribal and forest.
- ❖ To upgrade quality of tribal products with a view to maximise unit value realisation.
- ❖ To export tribal products.
- ❖ To provide full range of services, including organisation and collection.
- ❖ Tribal produce, scientific exploitation of forest products, storage.
- ❖ Transportation, marketing and exports.

4. Indian Farmers Fertilisers Co-Operative Limited

Mission:

Is "to enable Indian farmers to prosper through timely supply of reliable, high quality agricultural inputs arid services in an environmentally sustainable manner and to undertake other activities to improve their welfare".

The objects of IFFCO shall be to promote the economic interest of its members by conducting in affairs in professional, democratic and autonomous manner through self help and mutual co-operation for under taking manufacture/ production/ development of chemical fertilisers: bio-fertilisers, petro chemicals, refining industrial chemicals and hydro-carbon, their inputs etc.

IFFCO may undertake one or more of the following activities which are indicative but not limited to:

- ❖ To set up plant or plants for manufacture of chemical fertilisers and allied products/bye products.
- ❖ To undertake production, processing and manufacture of insecticides, pesticides, seeds, agricultural machinery etc.
- ❖ To acquire, establish, construct, provide and maintain and administer factories, townships, estates, channels, water reservoirs, storage sheds and accommodation of all descriptions for facilitating the business of IFFCO.
- ❖ To act as warehousing agency under the warehousing Act and own and construct its own go-downs or hire go-downs for the storage of fertiliser and other goods.

- To set up storage units for storing fertilisers.
- To maintain transport units of its own.
- To promote and organise other co-operative societies in the field of manufacturing.
- To subscribe to the shares of co-operative and other institutions.
- To take up such other activities which are incidental and conducive to the agriculture and rural development.
- To set up agricultural farms by purchasing, acquiring and taking on lease of land from Government institution or private agency for research and development of agriculture.

5. Indian Farm Forestry Development Co-Operative

It is a multi-state co-operative society promoted by IFFCO, has been implementing afforestation projects in UP, Rajasthan and Madhya Pradesh. The society has been floated under contribution agreement signed between IFFCO and India-Canada Environment Facility (ICEF).

Development of Primary Farm Forestry Co-operative Societies (PFFCS) is an important activity undertaken towards afforestation of waste lands. High participation of women is an important feature of the IFFDC.

6. Co-operative Rural Development Trust

IFFCO promoted Co-operative Rural Development Trust (CORDET) in the year 1979 to provide education and training to farmers on various aspects of crop production, horticulture, animal husbandry, farm machinery etc.

7. Krishak Bharati Co-Operative Limited

It is a premier co-operative society for manufacture of fertilizer, registered under Multi-state Co-operative Societies Act 1985, was promoted by (lie Government of India, IFFCO, NCDC and other agricultural I II operative societies spread all over the country.

KRIBHCO has set up a fertiliser complex to manufacture Urea. Ammonia and Bio-fertilizers at Hazira in the state of Gujarat, on the bank of river Tapti, 15 kms from Surat City on Surat-Hazira stale highway.

Objectives:

- To increase the urea installed capacity, maintaining its market share.
- To ensure optimum utilisation of existing plant and machinery.
- To diversify into other core sectors like power, LNG terminal/port, chemicals etc

8. National Bank for Agriculture and Rural Development

- NABARD serves as an apex financing agency for the institutions providing investment and production credit for promoting the various developmental activities in rural areas.
- Takes measures towards institution building for improving absorptive capacity of the credit delivery system including monitoring, formulation of rehabilitation schemes, restructuring of credit institutions, training of personnel etc.
- Co-ordinates the rural financing activities of all institutions engaged in developmental work at the field level and maintains liaison with Government of India, State Governments, Reserve Bank of India (RBI) and other national level institutions concerned with policy formulation.

9.3 Role of Cooperatives in Improving Livelihood of Farmers

9.3.1 Introduction

Cooperative society is an organization of group of people with collective responsibilities and thoughts for the development of needy, especially under privileged. "A cooperative is an autonomous association of persons united voluntarily to meet their common economic, social and cultural needs and aspirations through a jointly-owned and democratically-controlled enterprise." It is based on the values of the development of agriculture, forestry, banking, credit, agro-processing, storage, marketing, dairy, fishing and housing and its network covers 85 per cent of rural households. It occupies a key position in agricultural development with support in resource and input use, harvesting of water resources, marketing channels, storage facilities, distribution channels, value addition, market information and a regular monitoring network system. Cooperatives are also engaged in economic activities like disbursement of credit, distribution of agricultural inputs (seeds, fertilizers, and agrochemicals), etc.

Cooperatives have inherent advantages in tackling the problems of poverty alleviation, food security and employment generation. It is considered to have immense potential to deliver goods and services in areas where both the public and private sector have failed. By and large, cooperatives (except a few large ones) are "local institutions", addressing "local needs", employing "local talent", and led by "local leaders". A situation is emerging wherein there is a possibility of people migrating back to rural areas, as the cooperatives exhibit vast potential for generating self-employment opportunities at grassroots.

Since its formal launch in 1904 in India, the Indian Cooperative Movement has played very important role in the Indian economy especially in the development of the agriculture and rural sectors combining the strengths of both public and private sector specially in supporting the small and marginal farmers and weaker sections.

Immediately after Independence in 1947, the Cooperative Sector was recognised by the government as a third economic sector to serve as a balancing factor between the private and public sectors. The government had even considered the cooperative sector as a 'Cooperative Commonwealth'. In this way cooperatives developed as a special sector combining the strengths of both the public and private sectors, especially serving the needs of small and marginal farmers and weaker groups.

9.3.2 Improving Livelihoods of Farmers

Normally, a livelihood is known by any set of economic activities through which a household meets its basic needs and earns some cash income. When carried out on a repetitive basis, these tend to become a "way of life". The vast majority of people make a living through self-employment or wage-employment in the informal sector. In broader way, the livelihood comprises of people, their capabilities and means of living including food, income and assets. The tangible assets are income and stores, while the intangible assets are claims and access. The livelihoods should be sustainable environmentally and socially. The livelihood of farmers in India largely depends on agriculture and allied agricultural activities and the agriculture depends on land, water, Agri-inputs, credit, market, Govt policies and knowledge base.

Presently, with about 40% of the population living in poverty, livelihood security for the rural poor farmers continues to be a cause of concern in India. As a majority of the rural population is dependent on agriculture for their livelihood, increasing agricultural production through active involvement of the poor and weaker sections of the society is the most appropriate solution.

Small and fragmented land holdings, heavy soil erosion resulting in depletion of soil productivity, inefficient use of water resources, out-dated agricultural production technologies, un-availability of agricultural credit and lack of infrastructure for post-harvest management and marketing of agricultural produce, are responsible for lower yield and income.

Livestock is another source of income for small farmers. However, over 75% of the animals are uneconomical due to severe genetic erosion, inadequate feeding and health care. With lower crop and livestock productivity, the employment opportunities in the farming and other related sectors are reduced further, leading to reduction in farm wages, seasonal employment, malnutrition and migration.

Water is a critical input for food production and better quality of life. Neglect of water resources will not only affect the agricultural production and employment opportunities, but also affect the supply of clean potable water. Scarcity of water leads to unemployment, ill-health and hardship for women. This vicious cycle can be broken through improvement in agricultural production.

It is necessary to address the above challenges, particularly to improve the productivity crop yields of low productive non-irrigated areas, owned by small and marginal farmers, which in turn can improve the purchasing power of the poor. This calls for a value chain development approach, where small farmers are supported for backward and forward integration to improve production and add value to the produce.

There are many initiatives across the country where small and marginal farmers and tribal families, have excelled in food production in spite of small holdings, poor quality soils, limited resources and poor access to modern technologies, when they were assisted for establishing backward and forward linkages.

9.3.3 Role

Farming Cooperative Societies

Development of farming sector was given major emphasis after independence for ensuring food and employment security in rural India. While various land reforms were enforced to ensure equitable distribution of natural resources, consolidation of small and fragmented holdings was considered necessary for improving agricultural production. Hence concept of Cooperative farming was mooted by the Planning Commission to pool the land owned by small farmers for joint management. The proposed approach was either to retain individual ownership of the land and lease to the cooperative or transfer land ownership to the Cooperative and collect shares worth the value of the land. The Cooperative could then pool the holdings for land improvement and intensive cultivation, using modern technologies. These farming cooperatives were supported with financial resources to develop land and water resources.

By 1955-56, there were over 1000 cooperative farming societies with 1.9 lakh members, mostly in Punjab, Bombay and Utter Pradesh states. The Government of India had provided Rs.50 lakhs for promoting these societies. By 1960-61, the number of cooperative farming societies increased to 6,325 with 3.05 lakh members. These cooperatives had brought 3.5 lakh ha under cultivation. The number of farming cooperatives increased to 9,473 in 1971, with 2.57 lakh members and 4.88 lakh ha. Out of these 5,070 were joint farming cooperatives and 4,403 were collective farming societies. The programme was intended to assist small and medium farmers to come together for joint farming operations.

Agricultural Processing Cooperatives

The first processing cooperative society was established in India for setting up of a ginning factory in 1917. Subsequently, cooperatives for sugar processing, paddy milling, groundnut decorticating, copra and oil seed crushing, processing of fruit, vegetables, tea and jute were established. These processing cooperatives with individual farmers, cooperative marketing societies and local service cooperatives as members are regulated by cooperative rules and by-laws.

Farmers in India have been growing sugarcane from time immemorial but the first sugar plant was established by a French group in Orissa in 1824. The first vacuum pan process sugar plant was set up in Bihar in 1904. By 1932, there were 31 sugar factories, all in the private sector, producing 1.5 lakh tons while the annual consumption was about 12 lakh tons. Hence India had to import sugar mainly from Indonesia. With favourable policies, there was a sudden spurt in the establishment of sugar factories. By 1933-34, there were 111 sugar factories with 4.6 lakh ton production, which further increased to 148 factories with 11 lakh ton capacity by 1940-41, all in the sub-tropical regions of North India.

Today, with 316 cooperative sugar factories, India is the second largest producer of sugar in the world and the cooperative sector is responsible for about 48 per cent of the total production. The sugar industry has a huge annual turnover of Rs. 55,000 crores, ensuring livelihood for over fifty million sugarcane growing families and a large mass of agricultural labourers. Presently, 7.5% of the rural population of India is dependent on the sugar industry.

Dairy cooperative is another success story in India. To prevent the exploitation of marginal milk producers by the private dairy, farmers in Gujarat formed the "Kaira District Cooperative Milk Producers' Union Ltd." in Anand. Under this Union, cooperatives were formed at every village to collect surplus quantities of milk from small producers and a modern dairy was established to process the milk locally. The products were sold under the brand name Amul since 1955. The success of this dairy soon spread across the neighbouring districts of Gujarat. To support these Dairy Federations and to expand the activities, the Gujarat Cooperative Milk Marketing Federation, an apex marketing body was set up in 1973.

Presently, 170 Milk Producers' Cooperative Unions and 15 State Cooperative Milk Marketing Federations are involved in milk processing. However, many of these dairy federations could not function successfully, in the true spirit of a people's movement, barring a few states. Nevertheless, the dairy cooperatives account for the major share of processed liquid milk marketed in the country. During 2004-05, there were 12.95 million members of dairy cooperatives who sold milk and milk products worth Rs.3320 crores. Dairy Cooperatives have been instrumental in transforming the rural economy on the lines of cooperative sugar factories.

Agricultural Marketing Societies

Establishment of marketing cooperatives were encouraged to provide marketing facilities to small farmers. The anticipated advantages were increase in bargaining strength of farmers, removal of intermediaries and direct interaction with consumers. There was also scope for availing credit and cheaper transport, storage facilities, grading and processing of agricultural produce to fetch better prices. The National Agricultural Cooperative Marketing Federation of India Ltd. (NAFED) was established in 1958 for promoting cooperative marketing of agricultural produce. NAFED has been procuring foodgrains, pulses, oilseeds, spices, cotton, tribal produce, jute products, eggs, fresh fruits and vegetables from farmers through its cooperative network in selected areas whenever farmers faced problems of marking their produce. The advantages of cooperative marketing were increased bargaining strength of farmers, direct dealing with consumers, credit availability, cheaper transport, storage, grading and processing facilities and market intelligence.

Indian Farmers' Fertiliser Cooperative Limited (IFFCO) was established in 1967 to produce and distribute fertilisers through cooperatives. Presently, over 40,000 cooperative societies are members of IFFCO. Initially, IFFCO commissioned an ammonia – urea complex at Kalol and the NPK/DAP plant at Kandla both in the state of Gujarat in 1975. The another ammonia – urea complex was set up at Phulpur in the state of Uttar Pradesh in 1981. The ammonia – urea unit at Aonla was commissioned in 1988. Moreover, IFFCO has acquired an NPK/DAP and Phosphoric acid fertiliser unit at Paradeep in Orissa in September 2005. The marketing of IFFCO's products is channelled through cooperative societies and institutional agencies in over 28 states and union territories of India. In term of plan nutrient, it holds about 20% share in nitrogen and 30% in phosphorus production of the country. During 2013-14, it produced total 75.86 lakh MT and supplied 96.42 lakh MT fertilisers for the country. The annual turnover of IFFCO for year 2013-14 was Rs 20,853.45 crores.

At IFFCO, the thirst for ever improving the services to farmers and member co-operatives is insatiable, commitment to quality is insurmountable and harnessing of mother earths' bounty to drive hunger away from India in an ecologically sustainable manner is the prime mission. All that IFFCO cherishes in exchange is an everlasting smile on the face of Indian Farmer who forms the moving spirit behind this mission.

Agricultural Service Cooperatives

Cooperatives play a very important role in disbursement of agricultural credit. These cooperatives have been aiming at increasing agricultural production through credit supply to agricultural producers, agricultural labourers, artisans, supply of agricultural inputs, arranging storage, marketing and processing of agricultural produce, arranging raw materials for industries and providing technical guidance, while promoting social and economic welfare. There have been cooperative land development banks, which provide long term loans to cultivators for land development and capital investments.

The Primary Agricultural Cooperative Credit Society at the village level is the base for many of these activities. They federate into Central Cooperative Bank at the district level and with apex bank at the State level. The loans advanced by the cooperative banks increased from Rs.55 crores in 1950-51 to Rs.48,203 crores in

2005-06. At the same time, outstanding loans also increased to a large extent from Rs.20 crores to Rs.39,996 crores.

Allied Agricultural Cooperatives

These cooperatives cover activities like dairy farming, poultry, piggery, etc. Among them, poultry was a major activity with over 4,876 cooperatives with 4.4 lakh members in 2003-04. However, most of the members were not poor. There were 14,339 fishery cooperatives in 2003-04, but many of suffered from inadequate financial and technical support and lack of coordination between production, storage and marketing. There were 4714 forest labour cooperatives with 74000 members with a business turnover of Rs.227.23 crores in 2003-04.

9.4 Significance of cooperatives in Indian agriculture

The cooperative sector in India is the largest in the world and it plays a pivotal role in employment generation, poverty alleviation and food security. According to an estimate there are more than 500,000 co-operative societies working to bring about a change in the country. As seen, role of cooperative sector is extremely important in agriculture sector, as it supplies agricultural credit and funds and has the potential to deliver goods and services in vital areas where state and private sectors have not been able to do very much. Cooperatives are playing a critical role in creating mass awareness too.

As per the latest figures more than 75 per cent of India's population lives in rural areas, majority of which depends on agriculture for its livelihood. Indian agriculture is typically characterized with small and medium sized holdings, large number of agricultural workers spread over a large area, low income generation and low savings as a result of which the farmers look upon agriculture not as business but only as a way of life. Green Revolution during mid-1960's increased food production through the introduction of improved seeds, fertilizers, irrigation, farm mechanization, etc. all of which involve substantial financial investment that was ensured through both public support as well as personal interest of farmers. The compelling requirements of national food security and economic well being of farming community require further increase in production and productivity per unit area which is bound to invite still higher investment in agriculture. However, the economic status of a large majority of our farming community is not financially sound to make such investments. The farmer requires credit which drags him to exploitation of money lenders who charge exorbitant rates of interest. The adoption of principles of 'Cooperation' is the only solution for the progress of this type of agriculture and elimination of rural indebtedness in the country.

Cooperation was introduced in India in the beginning of twentieth century primarily for the supply of credit to agriculture. The basic objective of the cooperative movement in India was to eradicate local money lenders who usually exploit the farmers by charging unreasonable rates of interest. Agricultural credit societies were established with the promulgation of Cooperative Credit Societies Act 1904. In the beginning only small societies covering generally one each village were considered to be most suitable for Indian conditions. But soon it was felt that on account of small business and limited funds these societies could not effectively serve the needs of members. The cooperative structure has thus suitably been expanded to strengthen the facility of credit in the form of large societies and cooperative banking especially through supply of inputs like seed, fertilizer, weedicides, pesticides, etc.

It is now obvious that apart from easy availability of credit, the adoption of mechanization and efficient marketing of agricultural produce is a serious handicap of Indian agriculture. The cooperative societies have great relevance to provide credit, to ensure availability of farm machinery on contract basis, consolidation of small farming units through contract farming and joint marketing of farm produce in distant markets to draw maximum economic gains from farming. On account of inadequate storage facility and financial liabilities no individual farmer can hold back his farm produce to wait for remunerative price of his produce.

Cooperative societies can help farmers with such a facility at affordable cost and ensure better economic returns. In the current phase of increasing relevance of agro-processing and value addition to agricultural produce, cooperation becomes the only hope to wean the farmers away from the practice of distress selling of precious farm produce so that higher income can be realized through sale of processed products. Simultaneously, the agroprocessing units will provide the additional benefit of gaining rural employment. In addition to such obvious or direct involvement in agriculture, the cooperative societies have the potential to bring around social change in rural areas.

A substantial proportion of farmers' credit needs are now met by cooperative societies. The co-operative societies have led to the use of better farming methods such as the use of improved seeds, fertilizers, etc. The marketing and processing societies have helped members to buy their requirements cheaply and sell their agricultural produce at acceptable prices. It has also provided good storage facilities to the farmers to wait for appropriate time for higher prices. The non-credit societies, like the housing co-operative societies, the consumers' co-operative societies, etc. have helped their members to raise their economic conditions and have saved them from the exploitation by powerful groups. Cooperatives have very well served in the past and have great relevance to extend support to agricultural production in terms of provision of credit, input delivery and processing and marketing of agricultural produce to ensure adequate returns to farmers.

9.5 Role of NCDC

The National Cooperative Development Corporation (NCDC) was established by an Act of Parliament in 1963 as a statutory Corporation under the Ministry of Agriculture & Farmers Welfare.

9.5.1 Functions

Planning, promoting and financing programmes for production, processing, marketing, storage, export and import of agricultural produce, food stuffs, certain other notified commodities e.g. fertilisers, insecticides, agricultural machinery, lac, soap, kerosene oil, textile, rubber etc., supply of consumer goods and collection, processing, marketing, storage and export of minor forest produce through cooperatives, besides income generating stream of activities such as poultry, dairy, fishery, sericulture, handloom etc.

NCDC Act has been further amended which will broad base the area of operation of the Corporation to assist different types of cooperatives and to expand its financial base. NCDC will now be able to finance projects in the rural industrial cooperative sectors and for certain notified services in rural areas like water conservation, irrigation and micro irrigation, agri-insurance, agro-credit, rural sanitation, animal health, etc.

Loans and grants are advanced to State Governments for financing primary and secondary level cooperative societies and direct to the national level and other societies having objects extending beyond one State. Now, the Corporation can also go in for direct funding of projects under its various schemes of assistance on fulfillment of stipulated conditions.

9.5.2 Finance and Financing

Internal accruals, market borrowings and allocations from Government of India including International assistance. Purposes for which assistance is provided

- Margin money to raise working capital finance (100% loan)
- Strengthening of share capital base of societies (100% loan)
- Working capital to regional/state level marketing federations (100% loan).
- Term loan for creation of infrastructural facilities like godowns, cold storages, equipment financing, purchase of transport vehicles, boats and other tangible assets
- Term and investment loan for establishment of new, modernisation/ expansion/rehabilitation/diversification of agro-processing industries.
- Subsidy for preparation of project reports/feasibility studies etc.

9.6 Role of NAFED

9.6.1 Introduction

The National Agricultural Co-operative Marketing Federation of India (NAFED) is an apex organisation of marketing cooperatives in the country. The National Agricultural Co-operative Marketing Federation (NAFED) was established in

October 1958. The State Level Marketing Federations and the National Co-operative Development Corporation are its members. The head office of NAFED is at Delhi, and its branch offices are located at Bombay, Calcutta and Chennai. In addition it has the representatives of the Government of India, NCUI, NCDC, NCCF, Bharat Krishak Samaj and the State Bank of India.

NAFED's area of operation extends to the whole country. It deals in procurement, processing, distribution, export and import of selected agricultural commodities. The NAFED is also the central nodal agency for undertaking price support operations for pulses and oilseeds and market intervention operations for other agricultural commodities.

9.6.2 Objectives

The objectives of the NAFED shall be to organize, promote and develop marketing, processing and storage of agricultural, horticultural and forest produce, distribution of agricultural machinery, implements and other inputs, undertake inter-state, import and export trade, wholesale or retail as the case may be and to act and assist for technical advice in agricultural, production for the promotion and the working of its members, partners, associates and cooperative marketing, processing and supply societies in India. In furtherance of these objectives, the NAFED may undertake one or more of the following activities:

- to facilitate, coordinate and promote the marketing and trading activities of the cooperative institutions, partners and associates in agricultural, other commodities, articles and goods;
- to undertake or promote on its own or on behalf of its member Institutions or the Government or Government Organizations, Inter-State and international trade and commerce and undertake, wherever necessary, sale, purchase, import, export and distribution of agricultural commodities, horticultural and forest produce, other articles and goods from various sources for pursuing its business activities and to act as the agency for canalization of export and import and interstate trade of agricultural and other commodities or articles under any scheme formulated by the Government of India or other Government agencies and to facilitate these activities, wherever necessary, to open branches/sub-offices and appoint agents at any place within the country or abroad;
- to undertake purchase, sale and supply of agricultural, marketing and processing requisites, such as manure, seeds, fertilizer, agricultural implements and machinery, packing machinery, construction requisites, processing machinery for agricultural commodities, forest produce, dairy, wool and other animal products;

- to act as warehouseman under the Warehousing Act and own and construct its own godowns and cold storages;
- to act as agent of any Government agency or cooperative institution, for the purchase, sale, storage and distribution of agricultural, horticultural, forest and animal husbandry produce, wool, agricultural requisites and other consumer goods;
- to act as insurance agent and to undertake all such work which is incidental to the same;
- to organize consultancy work in various fields for the benefit of the cooperative institutions in general and for its members in particular;
- to undertake manufacture of agricultural, machinery and implements, processing, packing, etc. and other production requisites and consumer articles by setting up manufacturing units either directly or in collaboration with or as a joint venture with any other agency, including import and distribution of spare-parts and components to up-keep of the machinery/ implements;
- to set up storage units for storing various commodities and goods, by itself or in collaboration with any other agency in India or abroad;
- to maintain transport units of its own or in collaboration with any other organization in India or abroad for movements of goods on land, sea, air and operate freight station, container depot and undertake other incidental activities;
- to collaborate with any international agency or a foreign body for development of cooperative marketing, processing and other activities for mutual advantage in India or abroad;
- to undertake marketing research and dissemination of market intelligence;
- to subscribe to the share capital and undertake business collaboration with cooperative institutions, public, joint and private sector enterprises, if and when considered necessary for fulfilling the objectives of NAFED;
- to arrange for the training of employees of marketing/ processing/supply cooperative societies;
- to establish processing units for processing of agricultural, horticultural and forest produce, wool and allied products;
- to undertake grading, packing, standardization, scientific treatment and process of agricultural produce and other articles;
- to acquire, take on lease or hire, lands, buildings, fixtures and vehicles and to sell, give on lease or hire them for the business of NAFED.
- to advance loans to its members and other cooperative institutions on the security of goods or otherwise;

- to guarantee loans or advances or give undertakings to any Society or Company in which the Federation has a shareholding or financial involvements as a promoter to be able to assist its development or expansion or for starting any industrial undertaking by such societies/companies;

9.6.3 Activities

The NAFED performs the following activities:

Internal Trade

NAFED is engaged in interstate trade in agricultural commodities, particularly foodgrains, pulses, oilseeds, cotton, jute, spices, fruits, vegetables and eggs with a view to assuring better prices to the producers.

Foreign Trade- Export and Import of Agricultural Commodities

The NAFED exports agricultural commodities, particularly onions (canalised), potatoes, ginger, garlic, nigerseed, sesamessed, gum, deoiled cake of groundnut, soyabean and cottonseed, fresh and processed fruits and vegetables, spices- black pepper, cardamom, turmeric, cuminseed, coriander seed; cereals-rice, barley, bajra, jowar and ragi and jute bags.

Imports

The NAFED also arranges for the imports of pulses, fresh fruits, dry fruits, nutmeg (jaiphal), wetdates and chicory seeds and inputs particularly fertilizers and machinery as and when asked to do so by the government.

Price support operations

Government of India have designated NAFED as the nodal agency for implementing the price support policy for oilseeds and coarse grains during Seventh Five Year Plan period (1985-90). NAFED is being appointed as the agency of the government to undertake support price purchases of commodities like groundnut, soyabean, mustard seed, gram, tur, moong, bajra, jowar, maize, barley, toria and sunflower seed.

Production and Marketing of Agricultural Inputs

NAFED helps the farmers by supplying them agricultural machinery like harvesting combines tractors, spare parts and such other inputs as bio-fertilizers. NAFED also imports some of the machine and spare parts from abroad to ensure timely availability of genuine spare parts at reasonable prices.

Promotional Activities

NAFED maintains expert staff, which conducts market studies, collects data and circulates the results among the members.

Developing Co-operative Marketing of Tribal Produce

A separate cell to develop the marketing of product of the tribal areas (minor forest products) having economic value has been set up with the assistance of NAFED. NAFED arranges market intelligence, establishes better system for auction of tribal produce and develops for other commodities like chilgoza, gum, karya etc. Keeping in view the importance of marketing of tribal produce, a separate Tribal Co-operative Marketing Development Federation (TRIFED) has been set up.

Setting of Scientific Storage System

NAFED has set up a cold storage alongwith an ice factory and a warehouse in Delhi. It has set up an onion warehousing complex at Nagapattanam.

Processing of Fruits and Vegetables

NAFED has set up a multi-commodity fruit and vegetables processing unit at Delhi and at Vellore (Tamil Nadu). The basic purpose is to develop the processing industry in co-operative sector in a major way so as to make fruits and vegetable marketing and processing to the advantage of the farmers.

NAFED also manufactures jute goods in joint venture with Konark Jute Limited-promoted by Industrial Development Corporation of Orissa.

Co-operative marketing and processing of Milk and Oilseeds

Milk

With a view to providing marketing support to the milk producers, a sound network of dairy cooperatives has taken shape in the country. The network consists of milk producers cooperative societies at the village level, District Milk Co-operative Unions at the district level, State Co-operative Dairy Federations at the state level and National Dairy Development Board (NDDB) at the national level.

Oilseeds

Oilseeds grower's co-operative societies at the village level and State Co-operative Oilseeds Growers Federations have been organised in many States. The nomenclature differs from State to State.

References

1. Ashwathappa Dr. K. (1996) Business Environment for Strategic Management, Himalaya Publishing House, Bombay.
2. Bagchi A.K.(1987) The Evolution & the SBI the roots 1806-1876, Parti, II, Oxford University Press Calcutta.
3. Bandyopadhyay (1996) : Rural Credit System-An Alternative Approach, S.B.I. Monthly Review Vol-35.6.
4. Bhatt N.S. (1994) : Farm Production Credit Use Among Small Holder Farmers in Rural Nigeria. Journal of Rural Development, Volume 12, No.3, May, 1993.
5. Chakaraborthy Bakim (1996): Maharashtra State Co-operative Bank An examination of it's performance. Indian Co-op. Rev., 6(1).
6. Chandra Enagandula & Others (1999): Performance of the Karimnagar District Central Co-operative Bank (KDCCB) in Andhra Pradesh. An Economic Analysis. Indian Co-op. Rev., 4 (2).
7. Chawla A.S. (1978) Nationalization 85 Growth of Indian Banking, Deep & Deep New Delhi.
8. Chopra Sanjeev (1998) : The Structure and Organization of the Agriculutural Credit in India : Releationship between Primary Societies, Appex Bodies, and CDFAs. Research Paper presented at the inter-Regional Workshop on the Role of Co-operative Apex Organization in a Competitive Environment at Johannesburg, South Africa, 13-17 July 1998, Internet Data.
9. Cooke Northcot (1863) the Rise Progress And Present Condition of Banking in India Presidency Banks Act
10. Dandekar M. N. (1988): Farm credit in India. Journal of Rural Development, Volume 7, No.2, March 1988.
11. Daniel Thorner's (1954) : The Agarian Hand Book for India, The Agricultural Situation in India cited in ICSR opcit
12. Datta and Sundharam (1973):Indian Economy S. Chand and Company New Dehli.

13. Desai Vasant (2000) Banking & Financial Systems, Himalaya Publishing House, Mumbai.
14. Desai, Vasant (1995), Banking And Financial System, Himalaya Publishing House, Mumbai, Page 48.
15. Dr. Gopal Lal Jain (1997): Rural Development, Mangal Deep Publications Jaipur.
16. Gilbert (1932), History, Principles & Prictice of Banking Skyes Edition London.
17. Government of Andhra Pradesh, (2011), Gazzettier of Nizamabad District, Publication Dividion, Andhra Pradesh.
18. Government of Andhra Pradesh, (2011), Socio-Economic Analysis, Nizamabad District, Publication Dividion, Andhra Pradesh.
19. Government of India (1930): The Royal Commission on Agriculture Report of the Royal Commission on Agriculture, Calcutta, And Central Publication Division Bureau.
20. Government of India (1931) Report of the Central Banking Enquiry Commission, Govt. Press New Delhi.
21. Government of India (1931): Report of the Central Banking Enquiry Committee.
22. Government of India (1949): The Rural Banking Enquiry Committee Report of Rural Banking Enquiry Committee, New Delhi.
23. Government of India(1961-62) : All-India Rural Debt and Investment Survey.
24. Reserve Bank of India (2009), Report of the High Level Committee to Review Lead Bank Scheme, Publication Division, RBI Publication, Mumbai.

Index

www.ingramcontent.com/pod-product-compliance
Ingram Content Group UK Ltd.
Pitfield, Milton Keynes, MK11 3LW, UK
UKHW020844290926
15470UKWH00004B/1538